THE MECHANICAL ENGINEER IN AMERICA, 1830–1910

THE MECHANICAL ENGINEER IN AMERICA,

1830-1910 Professional Cultures in Conflict

by Monte A. Calvert

The Johns Hopkins Press, Baltimore

Copyright © 1967 by The Johns Hopkins Press
Baltimore, Maryland 21218
Printed in the United States of America
Library of Congress Catalog Card Number 66–26683

To My Parents, Bill and Evelyn Calvert

PREFACE

The inspiration for this study came from a series of conversations with Professor Samuel Haber, who led me to see the paradoxical character and actions of Frederick W. Taylor, and from Professor Samuel P. Hays, who introduced me to the concepts and practice of historical social analysis. In the research and writing, my ideas were sharpened by long and fruitful talks with Professor Robert Avery, my tutor in sociological thought. Professors Hays, Avery, William Stanton, and David Montgomery, and my Smithsonian colleague John H. White, Jr., all read the manuscript and gave helpful and constructive criticism.

My research was facilitated by the helpfulness of the staffs of the University of Pittsburgh Library and the Carnegie Library of Pittsburgh; the Library of Congress, the Patent Office Library, the Navy Department Library, the Museum of History and Technology Library of the Smithsonian Institution, in Washington, D.C.; the Cornell University Library in Ithaca, New York; the Engineering Societies Library in New York City; and the Library of the American Philosophical Society in Philadelphia, Pennsylvania. Jack S. Goodwin of the Museum of History and Technology Library provided valuable bibliographical assistance. Important manuscript collections and able staff assistance were found in the National Archives, the Cornell University Archives, the American Philosophical Society, the American Society of Mechanical Engineers, and the Stevens Institute of Technology Library.

A special debt is owed to my wife, Gertrude Baker Calvert, whose research assistance, editorial and conceptual criticism, and typing skill were equally as important as her encouragement in the dark hours that inevitably accompany such an enterprise.

Monte A. Calvert

CONTENTS

Key to Abbreviations

ASME = American Society of Mechanical Engineers
ASCE = American Society of Civil Engineers
AIME = American Institute of Mining Engineers
AIEE = American Institute of Electrical Engineers
AES = Association of Engineering Societies
WSE = Western Society of Engineers

LIST OF ILLUSTRATIONS

*All original photos or copies, unless otherwise
noted, are courtesy of the Smithsonian Institution.*

INTRODUCTION

Historical social analysis is something of an orphan. It is ignored by many orthodox historians, who secretly think of themselves as humanists, because it sounds like social science, particularly like sociology (a discipline paralyzed, they imagine, by jargon and statistics). Sociologists themselves are seldom willing to investigate historical phenomena primarily because comparable sets of data either do not exist or are hopelessly incomplete. In recent years a fruitful dialogue has developed between the two disciplines, which promises to enlarge the horizons of each; this movement may yet founder on the problem of finding comparable data.

When one approaches a problem such as the professionalization of the American mechanical engineer, he immediately is concerned with the question of social and occupational mobility, a concern complicated by the obscurity which surrounds the economic and social background of most engineers. Nor is it easy to use effectively the findings of other scholars regarding social mobility. An example is the problem of the Horatio Alger genre of myths about American society in the nineteenth century. Recently these myths have been under intensive analysis and resulting attack by sociologists, historians, and economists. The thesis presented here concerning the American mechanical engineer can better be placed in perspective by looking at some of the findings of these researchers. For example, they have asked: does a high general rate of social and economic mobility in a society prove that the leadership and membership of a particular occupation or profession of high social status did in fact come from a lower social and economic stream?

Consider the conclusions of Seymour M. Lipset and Reinhard Bendix regarding social mobility rates in industrial society. They found that social mobility tended to be high in any society which was in a period of rapid industrialization, thus laying to rest the ghost of the Horatio Alger success story's uniqueness to the American scene. The growing demand for skilled managerial ability prevents the permanent establishment of elites and, consequently, of rigid upper-class structures. They do recognize, however, that the social origins of professional groups exhibit national differences and may differ in mobility from such groups as business elites. Lipset and Bendix describe the findings of a series of studies of the American business elite which indicate that, since 1801 at least, the majority of prominent businessmen have come from families already well established economically.[1] The enabling circumstances such as family connections, higher education, and financial aid, to which this phenomenon is attributed, are summed up in the common phrase, "them that has, gets."

E. Digby Baltzell studied the Philadelphia business aristocracy, both upper-class and elite segments. He found an upper-class structure centered on past and present business elites; it was the only class with a self-conscious existence. Abjuring politics as beneath it, this class had its greatest influence between 1812 and 1940, a period well encompassing the period of professionalization of mechanical engineering.[2]

Starting with these two propositions, that the business elite tended to be self-perpetuating in nineteenth-century America, despite high social mobility, and that professional groups tend to have different mobility characteristics than do other groups in a society, one can conclude that the American mechanical engineer, with one foot in each group, may have a quite different origin than other professions and occupations.

The task of systematically investigating these assumptions in relation to the American mechanical engineer is complicated by the difficulty of assembling a representative group of mechanical engineers since sufficient biographical information is available for only a very few. These few are obviously the most successful, and usually the biographical information is incomplete on the matter of social origin. It is primarily through a reconstruction

---

[1] *Social Mobility in Industrial Society* (Berkeley: University of California Press, 1959), pp. 11–143.

[2] *Philadelphia Gentlemen* (Glencoe, Ill.: Free Press, 1958), pp. 3–130.

of complex interrelationships between individuals who worked in a few significant machine shops that one can begin to see the outlines of the American mechanical engineering elite and the social source of its development.

Professionalization, another focus of this study, is an el e concept to define since it has in our society great status v. ,ue and a normative function. For that reason no attempt will be made to determine whether or when the American mechanical engineer became a professional. Nor is it necessary to engage in an extended discussion of what constitutes professional status. Sociologists and historians disagree on the specific attributes of professionalization, partly, no doubt, because they have studied particular occupations in the terms appropriate to those occupations. From the welter of definitions of professionalization, I have extracted those attributes and activities which most sociologists, historians, and the general lay public in the United States would accept as factors contributing to a professional image for an occupation. This definition is presented only as a guide to the uninitiated and is not in any way definitive. It is based to a considerable extent on the model of the older, established professions of law, medicine, and the clergy, simply because most definitions used by social scientists today are based largely on these models. It is also biased in favor of the Anglo-American tradition. In short, the definition is not to be used in a normative sense, but more as a guide to what traditionally has been regarded as professionalization.[3]

The characteristic attributes are as follows:

1. The development by an occupation of a systematic, techni-

---

[3] See particularly Howard S. Becker and James Carper, "The Element, of Identification with an Occupation," American Sociological Review, XXI (June, 1956), 341–48; William J. Goode, "Community within a Community: The Professions," ibid., XXII (April, 1957), 194–200; Logan Wilson, The Academic Man (New York: Oxford University Press, 1942); Sigmund Nosow and William H. Form (eds.), Man, Work and Society (New York: Basic Books, 1962); Theodore Caplow, The Sociology of Work (Minneapolis: University of Minnesota Press, 1954); Harold L. Wilensky, "The Professionalization of Everyone," American Journal of Sociology, LXX (September, 1964), 137–58; William J. Goode, "Encroachment, Charlantanism, and the Emerging Profession: Psychology, Sociology, and Medicine," American Sociological Review, XXV (December, 1960), 902–14; William Kornhauser, Scientists in Industry (Berkeley: University of California Press, 1962); For an approach to the traditional professions in America before 1850, see Daniel H. Calhoun, Professional Lives in America (Cambridge, Mass.: M.I.T. Press, 1965).

cal knowledge base of an optimum size combining both theory and practice.

2. Recognition of the need for prolonged and specialized training for competence in the field and the development of educational institutions to effect this.

3. Recognition of the need (by those established in the occupation) for the socialization and control of both young and old practitioners. This generally includes a concern for use of titles, license to practice, and other forms of self-regulation.

4. A service orientation including a sense of responsibility to society (for competence in their area of public trust), a dedication to international scientific knowledge, and a limitation of self-interest.

5. The development of self-consciousness about social status and role by substantial and organized segments of the occupation and the establishment of professional associations.

6. The predominance of colleague over client orientation among practitioners, including, frequently, a code of ethics. Colleague orientation dictates a course of action on ethical questions which will be acceptable to one's peers—other members of the same profession. Client orientation refers to exclusive concern with pleasing the client, customer, or boss, as the case may be.

Defining the words "engineer" and "mechanical engineer" is an easier task than dealing with the mystique of professionalization. For the purposes of this study the mechanical engineer is defined as an individual concerned with the design, supervision of construction, and installation of basic machinery or power plants (such as machine tools or steam engines), or as one who creatively guides or shapes the technology of machine production. This is obviously a definition by functional role or by occupation. In addition, however, one must consider how the orientation of some practitioners led to interest in creating a separate profession. Certain occupations and certain technical problems created the conditions for the development of professionalism among mechanical engineers, and it is in these that one must seek the source of the phenomenon.

This book will analyze the process of professionalization as it occurred among American mechanical engineers from 1830 to 1910. Although mechanical engineering work was done much earlier than 1830 in this country, by that date one could identify a few individuals who were acting as mechanical engineers and held occupational and social status as such. Many reasons can be cited for using 1910 as a cut-off date: that was the first year mechanical engineering was listed as a separate occupation in the United States Census; after 1910 it is difficult to speak of the professionalizing activities of mechanical as distinct from other types of engineers; the end of the first decade of the twentieth century saw the establishment of sections and subprofessions within the American Society of Mechanical Engineers and the beginnings of mass growth by that technical organization.

Two major aspects of the professionalizing process will be dealt with: who pursued this professionalism and why, and who resisted it and why. I believe that from a detailed and analytical investigation of the nature of the conflict between the friends and foes of professionalism in one occupational group may come a better understanding of change in American urban-industrial society. This study will not try to evaluate the ultimate success or failure of mechanical engineers in reaching professional status. Rather it is a study of change and process with the focus on mechanical engineers and on their attitudes toward professional behavior and status.

The development of mechanical engineering will be dealt with in three parts: Part I, "Origins," will discuss the social, occupational, and cultural milieu from which the mechanical engineer emerged and his specialization and differentiation within the industrial world; Part II, "Internal Development," will trace the development of mechanical engineering education and professional associations as formal influences on professionalism, the changing role and status of mechanical engineers as informal influences, and will explore attitudes toward rationalization. Part III, "External Relationships," will investigate how the origins and internal development of mechanical engineering affected its practitioners in their relationships with nonprofessionals in mechanical occupations, with other professional engineers, with business and industry, and with the public. Special emphasis will be placed on the problems of naval engineers regarding status and on the development of scientific management.

My central theme is the identification and analysis of two groups, one of which favored, the other of which opposed, developments in engineering which would ordinarily be regarded as professional behavior. I believe that any attempt to truly understand strivings for professional status must transcend a mere effort to determine that a particular occupation did or did not become a profession by a certain specified date. In this case, the process is far more important than the end result.

PART I   ORIGINS

THE MECHANIC

Workman, artisan and dreamer,
His to puzzle and to plan;
Toiler, prophet, thinker, schemer,
For the betterment of man;
Through the turmoil and the panic,
Through the smoke and grime and murk
Looms the calm and wise mechanic,
Master of the world of work!

From The American Machinist
By Berton Braley

CHAPTER 1   THE AMERICAN MACHINE SHOP

At least since the beginning of the nineteenth century, Americans have had an inordinate love for the technique and practice of the mechanical arts. Much of this love may have been stimulated by the belief that only through the application of new sources of power, means of production, and modes of transportation would it be possible to subdue and populate the American continent. From 1790 to 1840 the United States was in preparation for an industrial take-off period in which the machine shop was to play a vital role. The shop became, in effect, the repository of all knowledge about mechanical problems because interest in mechanical theory was not as great as interest in the solving of practical problems.

Since the machine shop (along with the railroad shop and naval engineering) was also a major source of mechanical engineers, it becomes a central institution in this study. I will apply a new term and concept, "shop culture," to the orientation, institutions, and traditions of the nineteenth-century American machine shop. These shops and the traditions associated with them shaped the attitudes of the new profession in the first half of the century and in the second half continued to dominate the thinking of the mechanical engineering elite. These shops were quite different from factories, even though they were partly called into being by the needs of the growing factories in the years following American independence.

The textile industry was a major stimulus to machine shop growth. The development between 1790 and 1850 of large, complex, and precise spinning and weaving machinery created a demand for machined metal parts, for repair service, and, of course, for designing and building the machines themselves. Many shops were initially adjuncts to the textile plants, such as the Saco-Lowell group in Massachusetts. These several shops began making textile machinery for the parent company but branched out from 1814 to 1824, selling machines to other factories. After 1825 they began making machine tools and even locomotives for outside sale. These companies, though often controlled in part by parent company officials, became increasingly independent after 1840–50 (a decade in which innovation in textile machinery had declined). By the time of the Civil War, only a few had survived as major manufacturers of steam engines, locomotives, or machine tools. Nevertheless, they were a major part of the machine shop world in the first half of the nineteenth century.[1]

By 1810, independent shops appeared, to make steam engines to power factories, later to make engines for steamships, and by the 1830's to build locomotives. Most of these engines were made on special order, as were the machine tools (lathes, drilling and boring machines, and planers first, milling machines, shapers, and grinding machines later) which were produced in small quantities between 1820 and 1850. This American, heavy machinery industry remained relatively unspecialized before 1860, frequently making all types of heavy and light machinery in the same shop. Specialization began to appear by the 1850's and accelerated greatly from 1870 to 1900. By 1910 it was very difficult for a firm to make everything in the machinery line and survive.[2]

If any major stimulus to the growth and ultimate specialization of the American metal-working shop can be found (excluding specific industries like textiles), it was the development and perfection of a new source of power, the steam engine, and its application to factory production and to both land and water transportation. Thus the river steamboat was a very early stimulus, followed by the ocean-going steamship and the railroad after 1830. The factory created a need for steam power, for a heavy shop industry in the early period, and later, from

[1] George Sweet Gibb, *The Saco-Lowell Shops* (Cambridge, Mass.: Harvard University Press, 1950), describes a highly versatile shop complex. Gibb illustrates (p. 192) the diversification of the shops between 1845 and 1865 by the statement that they made everything from a jew's harp to locomotives. The story of a more specialized firm is found in Thomas R. Navin, *The Whitin Machine Works since 1831* (Cambridge, Mass.: Harvard University Press, 1950).

[2] Joseph W. Roe, *James Hartness* (New York: American Society of Mechanical Engineers, 1937), p. 37; Joseph W. Roe, *English and American Tool Builders* (New Haven, Conn.: Yale University Press, 1916); Wayne G. Broehl, Jr., *Precision Valley* (Englewood Cliffs, N.J.: Prentice-Hall, Inc., 1959), pp. 3–18, Nathan Rosenberg, "Technological Change in the Machine Tool Industry, 1840–1910," *Journal of Economic History*, XXIII (1963), 414–46, has interesting insights from the economist's point of view.

1860 to 1910, for precise, metal-working machine tools for use in mass production. If power dominated the machine shop concern in the first half of the nineteenth century, tools were a major rival in the second half.[3]

Machine shops came in many sizes and types. A small "job" shop might have been a one-room affair, doing repair work and building small steam engines and tools on special order. Such a shop might have had several lathes and perhaps a small planer after 1840. Many of these small shops grew up along the Erie and other canals and by the second half of the nineteenth century could even be found in small villages in the industrial Northeast. If a foundry was added, the enterprise often expanded to take on larger work. Some of these operations became quite large, although most did not without some specialization.[4]

The large, relatively specialized machine shop came into its own in the 1840 to 1860 period. Dozens of lathes of all types filled the floor of the shop, complemented by several planers and newer types of machine tools such as shapers, milling machines, and grinding machines. A separate erecting and assembling room contained locomotives, large machine tools, or steam engines in the process of completion. Highly specialized metal-working operations such as tin, copper, and brass smithing might have been contracted out but increasingly in this period were drawn into the shop complex itself. The foundry was large and well equipped. In a drafting room, plans were drawn for each project. A new phenomenon was the office, where books were kept and the operation directed by an engineer-owner and perhaps several assistants. This was the American machine shop at its height as an innovative institution.[5]

By 1870, factories producing heavy machinery appeared. As

[3] The rising importance of the machine tool trade can be seen in L. T. C. Rolt, *A Short History of Machine Tools* (Cambridge, Mass.: M.I.T. Press, 1965), pp. 137–77.

[4] Good descriptions of this type of shop can be found in Bruce Sinclair, "Delaware Industries: A Survey, 1820–1860" (unpublished research report, Eleutherian Mills-Hagley Foundation, December, 1958); its existence in the later period is documented by Fred H. Colvin in his *Sixty Years with Men and Machines* (New York: McGraw-Hill, 1947), pp. 15–17.

[5] J. Leander Bishop, *A History of American Manufactures* (2 vols.; Philadelphia: E. Young and Co., 1861), though not always accurate in details, can be consulted for a general impression of the physical plant and activities of such firms.

competition increased, as designs were perfected and standardized, it became more profitable to produce one type of steam engine or a line of standardized machine tools. As the machine shop became like the factory, it lost the uniqueness which gave it its importance as a source of mechanical engineering talent. Actually, all of these forms existed at the same time; however, each appears to have dominated a different period. Certain characteristics of the second type, the semispecialized, larger machine shop operation, typified in some degree its predecessor, the small "job" shop, and to a much lesser degree its successor, the heavy machinery-producing factory.

Most of the early shops worked almost entirely on special order rather than for a broad, competitive market requiring mass production and standardization. This was true partly because the designs of the products made, steam engines, machine tools, and locomotives, had not yet fully evolved. The shop frequently was an experimental laboratory which developed and perfected industrial and mechanical processes and equipment. In this role, shops developed better, more versatile machines than their customers required. At least this was true of the better firms that provided the leadership for what I have called shop culture.[6]

Competition was limited, partly because the owners of machine shops had close business and social relationships with their customers. Besides friendship, competition and the vigorously entrepreneurial spirit were further limited by customers' faith in a particular shop's ability to solve their mechanical problems and the machine shop personnel's vast knowledge of, and experience in, dealing with specialized problems. Furthermore, there was no contact with the nonindustrial customer

6 Colvin, Sixty Years, p. 94, describes the fruitless attempts to break a Sellers car-axle lathe by machinists who were promised a new one when it was worn out. The innovative role of these shops and the relation of innovation to risk are discussed in W. Paul Strassman, Risk and Technological Innovation (Ithaca, N.Y.: Cornell University Press, 1959), pp. 116–57. Studying the later period, and dealing also with light metal-fabricating shops, James H. Soltow, in his monograph "Origins of Small Business Metal Fabricators and Machinery Makers in New England, 1890–1957," American Philosophical Society, Transactions, LV (Philadelphia, 1965), does not find a strong pattern of innovation.

The majority of firms in any field are always, of course, imitative. My colleague John H. White, Jr., has pointed out to me that locomotive manufacturers were very loath to change or embellish designs that had proved to be workable, reliable, and cheap to produce.

since the machine shop's work was limited to supplying machines, advice, designs, layouts of plants, and repair services on machine tools and specialized mass-production machines.

One might expect an industrial institution whose capital consisted of know-how to be unwilling to share information with other shops. This was not the case. One function of shop culture was the sharing of information, which helped make it a preprofessional institution performing many of the functions taken over by the professional association and the school in the twentieth century. Machine shops exchanged information (particularly in such machine-tool centers as Philadelphia) to an extent unthinkable in competitive industries which sold to the general consumer on the open market. A popular feature in the early mechanical engineering periodicals was the section variously termed "shop kinks" or "shop hints," the purpose of which was to share every new technique, idea, and skill developed in one shop with all practitioners. Even though such ideas were mostly not patentable, once put into practice they might have increased the efficiency of competitors' shops; sharing them carried a certain risk. All this implied the conception of a vast, mutually owned store of knowledge and experience closely akin to a body of scientific knowledge.[7]

Another professional component of shop culture was the process by which the engineer and machinist became socialized.

---

[7] Eugene Ferguson suggests, but does not document, this phenomenon in "On the Origin and Development of American Mechanical Know-How," *Midcontinent American Studies Journal*, III (Fall, 1962), 15; many examples, however, can be found in his recent edition of *Early Engineering Reminiscences (1815–1840) of George Escol Sellers* (Washington, D.C.: Smithsonian Institution, 1965); see also the candid discussions of shop problems which occurred in the meetings of the American Society of Mechanical Engineers, *Transactions* (hereafter cited as ASME, *Trans.*), 1880–1910; evidence of the sharing of knowledge can be found in manuscript sources as well, such as letters of Alexander Lyman Holley to Charles Ridgely, January 9 and December 14, 1876, and to C. E. Emery, E. D. Leavitt, and George H. Corliss, December 7, 1878, in Holley Letterbooks (microfilms in the possession of the Smithsonian Institution); and in letter of J. F. Holloway to Coleman Sellers, March 27, 1880, and of John C. Hoadley to Coleman Sellers, June 21, 1852, in the Peale-Sellers Papers (American Philosophical Society, Philadelphia, Pennsylvania). Truly extraordinary are the actual plans and specifications for currently produced machinery published for all to learn from, such as G. Wiessenborn, *American Engineering, Illustrated By Large and Detailed Engravings Embracing Various Branches of Mechanical Art, Stationary, Marine, River Boat, Screw Propeller, Locomotive, Pumping and Steam Fire Engines, Rolling and Sugar Mills, Tools, and Iron Bridges, of the Newest and Most Approved Construction* (New York: G. Wiessenborn, 1861). For the role of *American Machinist* in the sharing process, see Colvin, *Sixty Years*, p. 41.

In the shop both men started as apprentices and worked side by side at the bench. Each gained the respect of the other and shared a common belief in the dignity of hand labor. The engineer who "graduated" from this system had an intimate knowledge of the technicians below him that was often lost in the later stages of industrialization. Here the engineer learned to establish relationships with his men without falling into the twin pitfalls of excessive familiarity or arrogant aloofness. Here he learned what was expected of him in lending dignity, sobriety, and dedication to the shop.

Theoretically the shop provided complete upward mobility from the lowliest apprentice to the engineers who ran the shops and the entrepreneurs who owned them—often the same person. In practice some men entered the shops fully expecting to become mechanical engineers because of their extensive social and business connections, and the routine of rising up the ladder step by step from apprenticeship had the air of ritual about it. The social origins of the mechanical engineers who came from the innovative machine shops in America indicate that upward mobility was not the primary operating factor in reaching the top levels of the mechanical occupations. The "room at the top" was limited and controlled by the men who were there in the interest of creating and preserving engineering as a gentlemanly profession.[8]

The claim of the mechanics' organizations of the period 1820–50 that the wealthy classes, the intelligent, refined, and "better" classes, did not apprentice their sons to trades and shops[9] is false when applied to the machine shop. For example, in 1839 J. Morton Poole started a machine shop to make textile equipment on the Brandywine Creek in Delaware near Wilmington. Poole trained in his shop a group of younger sons of the

---

[8] If this seems contradictory, a letter from the William Sellers Company in Philadelphia to Coleman Sellers (a cousin of the firm owner then working for Niles and Co. in Cincinnati), December 22, 1855, in Peale-Sellers Papers, both clears and muddies the waters: "We would add that we have felt more hesitation in offering this position [foremanship of the manufacturing division] to you from the fact of your being a relative, and the unpleasant duty that would devolve upon us in case you should not be able to assume the position . . . the same thing would on the other hand add to the pleasure of our intercourse should we upon trial find that we suited each other."

[9] For example, see "Condition of the Mechanics," New York State Mechanic (hereafter cited as N.Y.S. Mechanic), January 8, 1842, p. 54.

Delaware upper-class Quaker aristocracy, many of whom went on to start their own machine shops and to repeat the process with a new group of young apprentices. One of Poole's apprentices was William Sellers, his nephew and the son of the wealthy Philadelphia family of John Sellers, Jr.[10]

Sellers' ancestry included two eighteenth-century observers of the Transit of Venus (the passing of Venus between the earth and the sun, a major astronomical event), John Sellers and William Poole, and two Sellerses had been members of the American Philosophical Society. Young William was not sent to college. At the age of fourteen, after finishing the course at a private school sponsored by his family, he was apprenticed to his uncle. At twenty-one he left Poole's employ to take charge of the shops of Fairbanks, Bancroft, and Company, of Providence, Rhode Island, manufacturing mill gearing and steam engines. It is not denying Sellers' great talent and ability (which he later amply demonstrated) to suggest that so rapid a rise required some collusion among members of a shop aristocracy which by that time existed in the industrial Northeast.[11]

Sellers left Providence a mere two years later to return to Philadelphia and set up shop on his own. Edward Bancroft, a partner in the Providence firm, soon joined him. Bancroft's sister-in-law was the sister of James Morton Poole, and there were other relationships among the families of Sellers, Poole, and Bancroft.[12] After Bancroft's death Sellers went on to a career of innovation in mechanical engineering, revolutionizing and systematizing the manufacture of heavy, specialized machine tools, pioneering in the manufacture of interchangeable

---

10 Sinclair, "Delaware Industries," pp. 37–38.

11 By the 1850's it extended to include some Ohio communities, notably Cleveland and Cincinnati. See Coleman Sellers to Edward Bancroft, March 19, 1852, and George Escol Sellers to Coleman Sellers, January 7, 1853, in Peale-Sellers Papers.

12 See Ferguson, *Engineering Reminiscences,* for a fascinating account of the engineering activities of the other branch of the Sellers family in Philadelphia, the business and social relationships of which included such variant persons as Jacob Perkins, Oliver Evans, Samuel V. Merrick, Joseph Saxton, Nicholas Biddle, Matthias Baldwin, William Mason, Rufus Tyler, Professor Robert Patterson, Dr. Thomas P. Jones, and Isaiah Lukens. In George Escol Sellers' class in the mechanical drawing school run by William Mason were John C. Trautwine, William Milnor Roberts, Solomon W. Roberts, and John Dahlgren, all destined for careers in engineering. Sellers went to Europe at the age of twenty-four to inspect mechanical processes; he gained access, partly through family connections, to the best English machine shops.

iron bridge parts, and sponsoring (with Enoch Clark, the socially prominent Philadelphia banker) the production of the first large quantity of structural steel made in the United States. His personal acquaintance with the Philadelphia business aristocracy gave him an immense competitive advantage in terms of knowledge of the needs and requirements of his industrial customers.[13]

Sellers in turn passed his mantle on to a group of socially prominent members of his own family and the families of friends. Coleman Sellers, a Sellers and Co. partner, later a professor of engineering and consulting engineer on the Niagara power project, was one, and J. Sellers Bancroft, engineer and later independent shop owner, was another. Perhaps the best known of Sellers' protégés was Frederick Winslow Taylor, youngest son of William Sellers' upper-class neighbors in Germantown. Taylor, somewhat against the will of his parents, who would have preferred that he follow his father into law, apprenticed himself to "Uncle" William. Sellers raised Taylor to shop superintendent of the Midvale Steel Works and supported, over the constant objections of many of his staff, Taylor's costly experiments into the permissible rates of cutting metals, time study, and other attempts to rationalize work. Both the younger Sellerses, Bancroft, and Taylor were innovative mechanical engineers and supported the American Society of Mechanical Engineers (ASME) as members and officers.[14]

While William Sellers was at the Fairbanks and Bancroft shop in Providence, another young engineer, George H. Corliss, was getting experience there also. Corliss was the son of a country physician, who was, however, widely noted for his surgical skill and who was able to send his son to a private

---

[13] J. Thomas Scharf and Thompson Westcott, *History of Philadelphia (1609–1884)* (3 vols.; Philadelphia: L. H. Everts & Co., 1884), III, 2263–64; see also Carl W. Mitman in *Dictionary of American Biography* (hereafter cited as *DAB*) s.v. "Sellers, William." Sellers exhibited some interest in his family's scientific history and once wrote to the assistant secretary of the American Philosophical Society for some information about it. Sellers to Ella M. Morrison, June 21, 1895, in Archives, American Philosophical Society.

[14] Biographical matter on Taylor can be found in Frank Barkley Copley, *Frederick W. Taylor, Father of Scientific Management* (2 vols.; New York: Harper and Brothers, 1923). The Frederick Winslow Taylor Collection, Stevens Institute of Technology Library, Hoboken, New Jersey (hereafter cited as Taylor Collection), contains valuable additional material.

academy at Castleton, Vermont. Corliss was enamored of the idea of making a practical sewing machine when he went to work for Fairbanks and Bancroft, but he was put instead on engine design. He later designed and began to manufacture the Corliss steam engine, which incorporated several major innovations and was, for a time, the standard for efficiency the world over. His Providence shop and factory became another training ground for young aristocrats interested in mechanical engineering,[15] one of whom was Alexander Lyman Holley.

Holley was born into a well-to-do Connecticut family in 1832. His father, a successful cutlery manufacturer, was also governor of the state. Young Alex was sent to Williams Academy at Stockbridge, Massachusetts, a college preparatory school where he distinguished himself by his wit and intelligence. Reluctant to go to Yale as his family wished, in 1850 Holley enrolled in the recently established scientific curriculum at Brown University in Providence, graduating as a civil engineer. He sought work in the nearby shops of Corliss, Nightingale and Company, then spent a year running a locomotive on the Stonington and Providence Railroad, worked as a draftsman for a New Jersey locomotive works, where he made the acquaintance of Zerah Colburn, and began a journalistic career. Over the next thirty years he visited Europe thirteen times, was on personal terms with engineers and scientists the world over, perfected the Bessemer process for American use, helped found the ASME, and served as a symbol of the gentleman engineer. Though his degree was in civil engineering, Holley was clearly a mechanical engineer in terms of his work and commitment.[16]

All of the men who shared the connections described above were alike in certain ways. All were white, Anglo-Saxon, Protestant, and were born and raised in the industrial Northeast. Most of them were reserved, dedicated men noted for their strict discipline and seriousness. These characteristics describe the

---

[15] Dwight Goddard, *Eminent Engineers* (New York: The Derry-Collard Co., 1906), pp. 111–20.

[16] *Ibid.*, pp. 123–31; William R. Raymond, "Alexander Lyman Holley—A Memorial Address," ASME, *Trans.*, IV (1882–83), 35–74; E. W. B. Canning, Letter to the Editor, *American Machinist* (hereafter cited as *Amer. Mach.*), March 18, 1882, p. 3; James C. Bayles, "Address at the Unveiling of a Portrait of Alexander L. Holley," ASME, *Trans.*, XII (1890–91), 1059–64.

mechanical engineering leadership in the nineteenth century and may perhaps also describe the membership of the ASME before 1900. A network of interconnections existed among the members of the mechanical engineering elite, a network which was related to social and economic class, but which was also responsive to the criterion of talent. It was possible to enter the top ranks of mechanical engineering on talent alone,[17] but it obviously helped to be a member of the club by birth.

Similar interrelationships of personnel and horizontal mobility can also be seen in the machine-tool firms of Brown and Sharpe and Pratt and Whitney; in the pump works of Henry R. Worthington; in the Cleveland instrument-making firm of Warner and Swazey; in the Baldwin Locomotive Works in Philadelphia; and in a relatively small group of shops devoted to innovation in steam power and machine tools. There were several reasons why apprenticing their sons to such shops appealed to upper-class and elite groups in American society. These were not large, impersonal factories where apprentices or workers performed rote tasks; instead they were experimental shops or even laboratories where innovation in relatively new arts and sciences was taking place. The smallness of these concerns (usually a few dozen, almost always less than one hundred employees) made it possible for personal relationships to be a factor in advancement of apprentices and draftsmen. The fact that most were partnerships and not corporations further increased this personal aspect of the American machine shop.

To those who could see the handwriting on the wall, the steam engine and the growth of machine culture were destined to become a major intellectual challenge of the nineteenth century. To those to whom the money-obsessed narrowness of the entrepreneur-speculator appeared distasteful, engineering offered an opportunity to be in on the ground floor of the changes that were remaking American life. The machine shop offered both intellectual challenge and the opportunity to make a respectable profit. In short, provided it could be given professional status as an occupation, mechanical engineering offered a respectable way for the old business upper class to retain a foothold in the

17 For example, John Fritz and John Edson Sweet, about whom more will be said later.

new industrial society without entrapment in large-scale, mass-production manufacturing. It provided a gentleman's occupation within an ungentlemanly industrial world. This is not to suggest that this was a conscious action or even that it had a manifest relationship to upper-class and elite entry into the occupation; it is rather to suggest that the social outlook of the members and particularly the leadership of the mechanical engineering profession had such a function, even though it may well have been latent.[18]

American mechanical engineering as a profession had its origin in the mechanic and designer occupations of the metal-working machine shop. This shop might have been on an ocean-going steamship, at the central yards of a railroad, a theoretical one conducted on the pages of a technical journal, or in a northeastern manufacturing town, where it serviced the needs of factory and basic industry with machine tools and steam engines. Wherever it was located it offered the element of shop culture, a sharing of ideas and practical techniques that transcended the nonprofessional, competitive business world. Shop culture, however, retained a basic orientation to the profit motive, which limited the degree of professionalism possible; nevertheless, it provided both the occupational structure and the orientation toward industrial problems that characterized the profession of mechanical engineering.

One of the most significant areas in the shop culture complex, after the machine-tool and engine-building shops, was the railroad shop, which grew out of the need to service and repair locomotives and equipment and to solve certain problems in the application of steam power to land transportation. By 1850, American railroad development had reached such proportions that enormous railroad shops were established, which required the administrative and technical talents of a first-rate mechanical engineer. Altoona, Pennsylvania, was founded in 1849 on the main line of the Pennsylvania Railroad for the express purpose of providing such a shop operation.

As early as 1838 a number of operating railroads had engi-

---

[18] For the differentiation between manifest and latent function, see Robert K. Merton, *Social Theory and Social Structure* (Glencoe, Ill.: Free Press, 1949), chap. 1.

neering administrators who went by such titles as superintendent of engines and machinery, superintendent of motive power, and sometimes master mechanic.[19] As the shop and maintenance operations grew, however, many jobs opened up which required strictly practical knowledge and strong muscles. Just as the term "engineer" was appropriated by the engine drivers, the term "master mechanic" was taken up by a wage-earning group of skilled and semiskilled machinists who formed a protective association which did not exhibit professional attributes. The term "superintendent," of motive power or of engines and machinery, however, referred to a salaried position (approximately $1,000–$2,000 per year in 1860)[20] and a management position with technical prerequisites. It became one of the major fields for the development of the professional mechanical engineer. Despite his contact with management this mechanical engineer always had one foot firmly in the shop, and the railway shop became a significant example of shop culture.

By the 1850's these railway mechanical engineers began to develop a self-consciousness about their relationship to the administration of the railroads. Some of these men and some other railroad officials believed that once a railroad was built the special problems connected with operating it lay mostly within the domain of the mechanical arts and the science of dynamics rather than within the competence of either the civil engineer or the lawyer, speculator, and capitalist. In the mid-fifties some correspondents of the *American Railroad Journal* expressed the opinion that, if left alone, the mechanical superintendents could run the railroads more efficiently than could the directors chosen "with reference to their ability to raise funds, and not for their practical skill in the economical workings of the roads." What was needed to bring railroads out of their doldrums were talented and experienced men who understood that railroad management was a science and who could introduce order into it.[21]

Some railroads, even as late as 1860, did not employ a person with mechanical engineering abilities in the superintendent

---

[19] See the classified advertisements in *American Railroad Journal* (hereafter cited as *Amer. Rr. Jour.*) for 1838.

[20] "Master Mechanics," *Engineer* (Philadelphia), September 6, 1860, p. 29.

[21] A Railroad Officer, "Railroad Management," *Amer. Rr. Jour.*, April 12, 1856, p. 229.

capacity. *Engineer*, a Philadelphia journal, criticized such a policy as depriving the railroad companies of the services of "the best mechanical engineers." No self-respecting mechanical engineer would take a position where he was subordinate to a superintendent ignorant of mechanical engineering. *Engineer* also felt that the salaries for mechanical superintendents ($1,000–$2,000 annually) should be raised to at least $3,000–$5,000 per year to attract the best men. In any case, mechanical engineers should not be placed under former express agents, stage drivers, and clerks.[22] The importance of this sort of injustice was magnified, according to *Engineer*, by the fact that "in the United States, railroad engineering is, and must for a time continue to be, the principal branch of the profession [mechanical engineering]."[23]

Professional activity was stimulated by the common set of problems which these men faced and the common orientation which they acquired in handling those problems. Although an occasional individual in the 1840's might refer to himself in advertisements as a civil and mechanical engineer, and a few brave souls might use the title "mechanical engineer," it was in the 1850's and in the supervision of railroad machinery that it first came into general usage. The orientation that drew these men together in seeking title and recognition was the attempt to use science in solving railway problems and the concern for a systematic and regulated method of dealing with such problems in the future.

Thomas D. Stetson, who signed himself M.E. (for mechanical engineer) and who was a mechanical superintendent for the Erie Railroad, developed such ideas in detail in an article in the *American Railroad Journal* entitled "The System of Monthly Reports for Employees." Writing in 1855, Stetson believed that many railroads were run with no plan at all, particularly with respect to the coordination of men and machinery. Not so with the Erie, Stetson claimed, where the superintendent "knows the capacity and daily use of every engine on the road; he has at hand the position of every car, the pounds of freight in each

22 "Master Mechanics," *Engineer* (Phila.).
23 Editorial, *Engineer* (Phila.), August 16, 1860, p. 5.

train, the average speed between two stations, the delay at each end and the freight handled, cars switched or train waited for. He knows the proportion of dead weight to useful load in every case." Stetson's article clearly indicated the necessity that the superintendent be a man with the technical qualifications to solve such a logistic puzzle.[24]

In recognition of the growing technical complexity of railroad operations, the *American Railroad Journal* inaugurated a mechanical engineering department in June, 1853, under the direction of Zerah Colburn. A brilliant, erratic, self-taught engineer, Colburn apparently never was a railway mechanical superintendent. But he devoted his career in technical journalism to identifying and trying to solve the problems faced by the railroad draftsman and designer, master mechanic, and superintendent of motive power and machinery. His principal mission was to induce them and their employers, the railroad directors, to use applied science in solving the problems of efficient railroad management.

By the beginning of 1854 Colburn had written a series of articles on such technical subjects as "On the Waste Heat of Locomotive Boilers," "Safety System of the New York and New Haven Railroad," "Improvement of the Locomotive," and "Mechanical and Financial Disadvantages of Grades upon Railroads."[25] Colburn advocated that the following considerations be taken into account when matching locomotives to the grades they worked: "diameter of cylinder, length of stroke, no. of drivers, diameter of drivers, number of trucks, weight on each back wheel, weight on each front wheel, whole weight, diameter of boiler, no. of tubes, diameter of tubes, length of tubes, tube surface, fire box surface, grate surface." He berated those engine builders who relied not upon science but upon "preference, based upon a primitive practice of engineering."[26]

24 October 27, 1855, pp. 675–76. The whole problem of the impact of the railroad on American business and technological life is documented in Alfred D. Chandler, Jr. (ed.), *The Railroads—The Nation's First Big Business* (New York: Harcourt, Brace, and World, 1965).

25 *Amer. Rr. Jour.*, January 7, 1854, pp. 13–14; *Amer. Rr. Jour.*, January 21, 1854, pp. 37–40; *Amer. Rr. Jour.*, August 12, 1854, p. 506; *Amer. Rr. Jour.*, January 28, 1854, pp. 50–52.

26 "Economical Working of Grades," *Amer. Rr. Jour.*, February 4, 1854, pp. 68–69.

Colburn made clear the creative and innovative role he expected the mechanical engineer to play in contrast to the mere
technical proficiency of the mechanic in a discussion of the
optimum size of locomotives: "To show *how* engines of a size,
greater than those in present use, can be made, is nothing very
difficult for any mechanic. I think, however, that I have shown
why they should be so made."[27] The above was written just
three weeks after Colburn had, for the first time in his professional life, signed himself "Zerah Colburn, *Mechanical Engineer*."[28] A week later he stopped working for the *Journal* and
simultaneously quit his job as superintendent of the New Jersey
Locomotive Works to take up technical journalism on a full-
time basis.

Railroad Advocate and the other short-lived journals Colburn
edited singly and in partnership with Alexander Lyman Holley,
from 1854 to 1857, hammered consistently at the unsystematic
methods of handling construction, machinery, and management
of American railroads. Holley, who was later one of the leaders
in the movement to form the ASME, was probably the instrumental member of the pair in securing funds from a group of
railroad presidents to finance a fact-finding trip to Europe. The
report on that trip furnished one basis for technical improvement of American railroads and became a classic in its own
time.[29]

Railroading created certain mechanical problems which were
best solved by men acting as mechanical engineers. Railroad
development provided the impetus for the introduction of the
term "mechanical engineer" into common practice in America
and stimulated the first serious examples of technical journalism
in this country in the area of mechanical engineering. It provided
employment for mechanical engineers outside the independent
machine shop. It is not surprising that railroad mechanical superintendents and technical journalists were among the founding
members of the ASME in 1880.

27 "The Economy of Railroads, as Affected by the Adaptation of Locomotive Power—
Addressed to the Railroad Interests of New England," *Amer. Rr. Jour.*, April 22, 1854,
p. 245.
28 *Amer. Rr. Jour.*, April 1, 1854, p. 193.
29 Raymond, "Holley," p. 55.

Pre-1855 machine shop in the Hall of Tools, Museum of History and Technology, Smithsonian Institution, Washington, D. C.

Naval engineering was another source of professional mechanical engineers and in some ways was very different from shop culture. It particularly illustrated, however, the social origins of the earliest mechanical engineers. When the United States first began experimenting seriously with steam vessels in the late 1830's and early 1840's, it was recognized that more than a mere engine driver would be needed to run the steam warship at sea. For diplomatic reasons warships frequently could not make port for repairs in wartime, and the proper functioning of all machinery in battle was a prerequisite to any hope of victory. Scientific engineers with a specialty in steam mechanics and design were needed; to attract the leaders in this field, the position had to be made attractive to gentlemen. As Abel P. Upshur wrote, while Secretary of the Navy in 1841: "The use of steam-vessels, in war, will render necessary a different order of scientific knowledge from that which has heretofore been required. If our navy should be increased by the addition of any considerable number of steam-vessels, engineers will form an important class of naval officers."[30]

The first assistant engineers, as they were termed, were appointed to serve in 1837 on the first regularly commissioned steam vessel in the United States Navy, the *Fulton*. Since some of these first appointees were men of considerable technical prestige and social status when they were hired, they were initially extended the privileges of the officers' wardroom even though they had no regular commissions.[31] Both Charles W. Copeland and Charles Haynes Haswell, members of this exclusive club, were recognized and respected designers of marine boilers and other equipment before they took positions as naval engineers. Thus was the position invested with professional navy social status even though many years passed before it achieved command rank. After leaving the Navy, Copeland set himself up as a "Consulting Steam Engineer." He was a founding member and early officer of the ASME some forty years later. Haswell stayed on as Chief Engineer of the Bureau of

---

[30] Quoted in Henry L. Burr, *Education in the Early Navy* (Philadelphia: n.p., 1939), p. 207.

[31] Frank M. Bennett, *The Steam Navy of the United States* (Pittsburgh: Warren and Co., 1896), p. 28.

19

Construction, Equipment, and Repair from 1844 to 1850 and laid the groundwork for the profession of naval engineering, which contributed both men and status to the young field of mechanical engineering.[32]

By 1842 the Secretary of the Navy could appoint one chief engineer, two first assistants, two second assistants, and three third assistants for each steamship of war.[33] Haswell, and the men who followed him as chief engineer in the 1850's, set high standards for these various positions for which they were allowed to make recommendations. Applicants for the engineer slots were given stiff examinations in steam engineering and in general mechanical engineering, covering such scientific subjects as mechanical powers, expansion of steam, strength of metals, resistance of solids through fluids, and geometrical figures. To pass them the applicants clearly needed and had, as an investigation of the actual exam papers shows, considerable knowledge of the basic principles and practice of mechanical engineering, mathematics, and physics.[34]

The Naval Academy at Annapolis, Maryland, which was established in 1845, was to some extent created in response to the needs of a modern steam navy. General science, not mechanical engineering, was a large part of the curriculum at the Academy; however, not until 1866 was a special course in steam engineering offered. Before the mid-sixties, the Naval Academy did not furnish men to fill the engineer positions on steam warships. Where then did these men come from, these "young men . . . of liberal education, intelligence, and natural mechanical ability"? The country needed, continued the writer of an 1838 article reprinted in the *American Railroad Journal*, many more competent steam engineers if she were to recapture the lead in marine engineering from Britain, whose *Great Western* was then winning transatlantic honors. The writer applauded the fact that "our government has already employed some of

32 For an appreciation of Copeland's work for the Navy, see Charles W. Copeland to Com. Charles Morris, Board of Navy Commanders, January 11, 27, and December 16, 1840, Entry 224, Record Group 45, Naval Records Collection of the Office of Naval Records and Library, National Archives. Hereafter records in the National Archives are indicated by the record group (RG) number, followed by the symbol NA.

33 Bennett, *Steam Navy*, p. 40.

34 Records and Minutes of Examining Boards, 1845–1856, Entry 156, RG 24, Naval Records, NA.

the members of this profession in the naval service, and many more ere long will be required for similar duties, effecting not only a further demand, but adding a most honorable feature as well as character to the profession."[35] Naval engineering did create the avenue to professional status for mechanical engineering that the young engineer-writer envisioned, but our concern for origins leads us back to the question of where these men came from.

Fortunately the examination papers of the applicants for the positions yield the answers. One of the questions asked related to previous experience and requested the applicant to specify shop experience, service on land engines, and service on boat engines. Coupled with the age of the applicants, this data gives a highly consistent picture of the typical applicant for naval engineering positions. In 1845, 27 individuals were examined for positions or promotions in the naval engineering service. They ranged in age from 17 to 51 years, the average being 28.3 years. Their average service for running boat engines was 3.5 years, and the average for running land steam engines was only 4.3 months, making a total of just under 4 years average engine-driver service. Service in machine shops, however, averaged 5.9 years, or half again as much.[36]

TABLE 1: Previous Experience of Naval Engineers

Year	Total Number of Individuals	Age Range (Years)	Mean Age (Years)	Mean Engine Driving Experience (Months)	Mean Shop Experience (Months)
1845	27	17–51	28.3	46.3	70.8
1855	17	20–26	22.1	9.5	38.4
186?	58	20–42	24.6	35.3	41.2

In 1855 there was an even higher ratio of shop to engine-driver service. Of 17 individuals, ranging in age from 20 to 26 years and averaging a much younger 22.1 years, average boat

35 A Young Engineer, "Remarks on Steam Navigation, Explosions, and Engineers," *Army and Navy Chronicle,* reprinted in *Amer. Rr. Jour.,* October 15, 1838, p. 260.
36 All material regarding examinations and applicants comes from Records and Minutes of Examining Boards, Engineers Corps, Entry 156, RG 24, NA.

engine service ran 1.7 months, average land engine service 7.8 months, making a total average of 9.5 months as engine drivers. By contrast, shop service experience amounted to a 3.2-year average, making a better than three-to-one ratio of shop over engine-driving experience.

A typical year in the early 1860's gives a quite different picture. Ranging in age from 20 to 42, with an average age of 24.6 years, the applicants averaged 10.7 months on land engines and 24.6 months on boat engines, for a total of 35.3 months of engine-driving experience. Shop service, however, only averaged 41.2 months, clearly showing a change in the origin of the applicants. What do these figures indicate?

First, they indicate that from the founding of the naval engineering specialty in the late thirties until 1860 most naval engineers came from the machine shop and very few had extensive experience in running engines. During the Civil War, however, the need for large numbers of engineers drove the Navy to lower the standards of the corps and appoint acting assistant engineers. That a lesser achievement and knowledge was expected of these men is indicated by the much more practical and simple examinations given them. The exams included penmanship, spelling, sketching, management of engines, pumps, boilers, and fires, safety valves, and other basic practical knowledge. This resulted in a higher rate of application on the part of those who had mostly engine-driver service.

The high age of the 1845 applicants indicates that the expansion of the naval engineering service in the 1840's presented an opportunity for a group of men who had extensive shop experience to find a professional niche for their talents. The naval engineering service was tapping a waiting reserve. The much lower average age of the 1855 group is perhaps partly explained by the fact that the depression of the mid-fifties put a group of younger men from the shops in unfavorable positions. The ratio of shop to engine-driver experience proves that these men were not basically engine drivers, but were men with significant experience in shop culture.

It was the shop that contributed the personnel to man the Naval Engineer Corps, men who were able to gain and retain more or less equal status with professional naval officers in the

1840's and 1850's. Naval engineering service provided an avenue to professional engineering status for the mechanical engineer similar to that provided by army engineering for civil engineering. Naval engineering provided the avenue, but shop culture provided the individuals and their orientation.

Metal-working machine and engine shops, railroad shops, and naval engineering were the three major contributors to the traditions of preprofessional activity in mechanical engineering. There were other occupational groups which might have been expected to contribute personnel or orientation to the new profession of mechanical engineering, such as civil engineers, millwrights, and engine drivers on land and sea. They did not, and not the least of the reasons why was that they failed to, or did not wish to, develop any permanent relationship with shop culture.

Civil engineering, defined in this country as the science of constructing canals, bridges, railroad beds, and aqueducts, might appear to have been a logical source for future mechanical engineers. Called into being as a profession by the canal- and railroad-building activities after 1815, civil engineers faced mechanical engineering problems in the building and maintaining of canals and railroads. They were called upon to do many types of theoretical and practical work which later were done by mechanical engineers. Civil engineers were, however, constrained by a considerable attachment to the idea that theirs was a gentleman's profession and that mental and manual labor did not mix. In addition, those who were more professionally oriented had a distinctly antientrepreneurial bias. Civil engineers wished to be regarded as consultants, artists, architects, but certainly not as businessmen. They were challenged by the antiprofessional and anti-intellectual attitude of the free-wheeling, highly mobile, entrepreneurial society of the America of the first half of the nineteenth century to justify their importance to a project in terms other than ownership.[37]

The civil engineers themselves had emerged from a previously established professional group, the military engineers, and the connection was by no means severed. Throughout the

[37] Daniel H. Calhoun, *The American Civil Engineer* (Cambridge, Mass.: M.I.T. Press, 1960), pp. 1–90.

first half of the nineteenth century the bulk of civil engineers were produced by the United States Military Academy at West Point. Thus the American civil engineer has been from the very first acutely self-conscious about his relationships to business, science, art, and society. This self-consciousness precluded any significant involvement with the mechanic arts.

In the second half of the nineteenth century few civil engineers became mechanical engineers. The educational institutions and professional associations of the civil engineers developed a disdain for mechanical engineers and other groups which formed professional engineering associations in the late nineteenth century, maintaining that the bulk of the membership in these newer occupations were not gentlemen in their sense of the word. Civil engineering contributed very little to mechanical engineering, except perhaps a professional model, and even that was difficult to follow because that model was based on the independent consultant, a work role seldom occupied by mechanical engineers. Clearly, they did not contribute either personnel or orientation to mechanical engineering, and they did not have any relationship to shop culture.

In spite of their high professional orientation, American civil engineers found it difficult to create a lasting professional association. This was partly because the nature of their work drew them to the far ends of the country and made it difficult or impossible for them to congregate in one place for meetings. Such was the case in 1839 when an attempt to create an American society of civil engineers aborted because, of the seventeen major civil engineers involved, only five were located in the Northeast (two in New York, two in Pennsylvania, and one in Connecticut), the remaining twelve being south of the Mason-Dixon line. Regardless of the success or failure of this early attempt at association, the proposed constitution made clear the basis for membership, with a separate category of associates which would include "architects, eminent machinists, and others, whose pursuits constitute branches of engineering, but who are not Engineers by profession."[38]

---

[38] Edward Miller, "Constitution Proposed for the American Society of Civil Engineers, with Proceedings in Reference to the Same, April, 1839," *Amer. Rr. Jour.*, April 15, 1839, p. 227; "Communications," *Amer. Rr. Jour.*, February 1, 1840, p. 82.

If the civil engineer, and by implication the military (army) engineer and the architect, had little concern or interest in the problems of mechanical practice, one group of constructors did have such interest—the millwrights. Originating in the application of basic mechanical functions such as the screw, the inclined plane, the crank, and the pulley, the millwright was essentially an artisan lacking any pretension to professional status. His specialty was the conversion of animal, water, and wind power into useful work through the construction of grist mills, water pumpers, and other types of mills to ease mankind's burden. His concern was with wheels, gears, and screws in motion coordinated to bring about useful mechanical action. This is, in essence, the work of the mechanical engineer.

Mechanical engineers themselves, even in the early period of professionalization, recognized a debt to the millwright. As early as 1860 an American engineering journal editorialized that, after civil engineering had fully established itself, "it was not long before the millwrights became mechanical engineers, or that mine adventurers made themselves known as mining engineers."[39] Thirty-five years later an old engineer attending a meeting of the ASME recalled that "the millwright of fifty years ago was the mechanical evolution of the preceding ages from the time of Archimedes, and was supposed to know everything pertaining to machinery and mills, from a watch movement to a fifty-foot water wheel."[40] Such reflection by engineers themselves does not serve as proof, however, since only a part of the mechanical engineering profession had their origin in millwright occupations. Several factors worked against any such simple transition.

The millwright's craft was bound up in the Eotechnic (to use Lewis Mumford's apt distinctions) or wood-centered culture. This presented definite problems of transition to the Paleotechnic or iron- and coal-centered society that emerged in the first half of the nineteenth century.[41] A few gifted persons did make the transition, such as Oliver Evans, who was both millwright and nascent mechanical engineer. Evans designed elabo-

---

39 *Engineer* (Phila.), August 16, 1860, p. 5. This journal was edited by Zerah Colburn.
40 Olin Scott, comment on Robert Allison, "The Old and the New," ASME, *Trans.*, XVI (1894–95), 742–61.
41 *Technics and Civilization* (New York: Harcourt, Brace, and Co., 1934), pp. 107–211.

rate automated grist mills, improved and developed the idea of the high-pressure steam engine, and constructed a working amphibious steam carriage at the turn of the nineteenth century. The very virtuosity of Evans' activities suggests something atypical, and the examples of millwrights becoming creative and innovative mechanical engineers are very few.

A second factor which prevented the average millwright from developing the outlook and occupation of the mechanical engineer was his transitory geographical situation. He was frequently a tramp, roaming about the country in search of work. This same geographical separation and transient nature, which prevented the civil engineers from effective organization but did not inhibit their professional sense, effectively quashed any possible professionalism among millwrights. There is no record that millwrights ever tried any type of associational activities. Although they were, in effect, replaced in function in the paleotechnic world by the installing or erecting engineers, a mechanical engineering specialty, millwrights as an occupational group played no effective part in developing the profession of mechanical engineering. They lacked, in fact, a permanent relationship to shop culture, which became the central reference point of American mechanical engineering.

The failure of several other occupational groups, none of which achieved any sort of permanent relationship with the shop, to become source groups for nascent mechanical engineering is further evidence of the power of the institution. The inventor, for example, although he did visualize and build mechanical creations, rarely worked under any sort of contract with manufacturing industry. He frequently invented something for which he felt there was a need, rather than something specifically requested by industry, and even when the need was clear, he operated as a selfish, independent entrepreneur and exhibited no interest in sharing his ideas with others. This was caused largely by the unusually well-conceived and well-run American patent system, which gave great importance to secrecy and unlimited possibilities for profit and loss. The inventor in America was placed in the position of being one against the world, a situation hardly likely to lead to professionalism.

As the use of the steam engine spread in the early nineteenth century, men had to be found to run the engines. This need created other possible sources of mechanical engineers in the engine-driver occupations: the locomotive engineer, the marine engineer, and the stationary engineer. It was immediately apparent to the owners of such engines that very little talent was required to keep them running. The result was that, as the steam engine's operation was perfected, the owners hired the cheapest labor possible to run them. This had the effect of forcing those engine drivers who thought themselves truly competent to band together, primarily for the purpose of securing state license laws which would prescribe the qualifications of the operating engineer in detail. The associations and organizations created by these groups sought to get license laws with such vigor that no other type of professional activity was pursued. In fact, an overriding concern with state or national license laws may not represent professional behavior since it relieves the occupational group of the responsibility for policing the practitioners. And, in fact, the engine drivers themselves were echoing the already loud cries of the public for regulation of steam boilers and those who tended them, prompted by vivid newspaper reporting of many disastrous explosions on land and on rivers, lakes, and oceans.[42]

Furthermore, the work which the engine drivers did was not conducive to the development of interest in broad theoretical questions, or even to a wide interest in all types of practical mechanical problems. Partly, this was related to the fact that none of these occupations had any direct connection (with the possible exception of the marine engineer on the larger ocean-going steamship) with shop culture. All were involved with the more or less isolated job of tending one engine, which precluded their coming into contact with a large variety of mechanical problems. Except in rare cases these occupations did not lead one into professional mechanical engineering.

[42] For a description of the National Protective Association of Practical Engineers Annual Convention, November, 1858, at Louisville, Kentucky, see "Locomotive Engineers," *Amer. Rr. Jour.*, November 28, 1857, p. 760.

CHAPTER 2 THE MECHANIC

Since the machine shop had an important place in the professionalization of the American mechanical engineer, one might suppose that the mechanics' organizations and magazines which flourished from 1820 to 1850 in the industrial Northeast were representative of shop culture. They were not, however, since the name "mechanic" was appropriated by the free artisan group (at least by the vocal elements of it). Thus the mechanics' movement had little or no direct relationship to the development of the professional mechanical engineer, except inasmuch as the movement did raise and try to answer certain basic questions of professionalism in general. And, of course, some of the mechanics were actually metal workers in machine shops.

Mechanics' organizations were active in the northeastern United States in the period from 1820 to about 1850, but two centers of such activity, Boston and the complex of shops in the Albany-Troy area at the head of the Erie Canal, are of particular interest. It appears that the leadership was composed of entrepreneurial-artisan types who either owned or expected to own their businesses and that membership consisted of apprentices and workmen in the shops, many of whom would have shared the same aspirations. Despite a predominantly middle-class outlook, these self-styled mechanics still worked with their hands, and this fact shaped their attitudes toward education, social status, professions, and their society.

One sample of the actual membership of these organizations gives a fairly good impression of the New York groups. A complete breakdown of the occupations of those attending an 1835 mechanics' convention on prison labor at Utica, New York, was printed in the *American Railroad Journal*. Of the ninety-eight self-styled mechanics who attended, only eighteen were engaged in occupations which might later have been the legitimate concern of the mechanical engineer. These eighteen broke down into: blacksmiths, four; tin plate workers, three; plane makers, three; watch makers, two; iron founders, two; locksmiths, one; brass founders, one; coppersmiths, one; and machinists, one. Of the remaining eighty, most would be classified as skilled tradesmen, such as cordwainers, tailors, printers, carpenters, cabinet makers, and many others.[1] If the associations themselves seemed to have a low percentage of persons engaged in mechanical occupations, they probably reflected the percentage of metal-working mechanics among the free artisan population.

In the case of the Boston associations, one must rely upon ideological evidence entirely. Most of them claimed to be associations of apprentices, which may have been true and thus may partly explain their lack of political action. The New York groups seem to have been associations of journeymen and perhaps even masters, who were older and thus more politically mature and aware of their substantive interests. None of this is meant to suggest that the European systems of craft guilds and carefully regulated apprenticeships had been introduced into the United States. They had not; yet many of the mechanics' organizations expressed interest in creating a legally defined and enforceable apprentice system.

Foremost among the objectives of the mechanics' organizations, clubs, institutes, and associations which were formed was "mutual encouragement and aid in the great enterprise of mental, moral, scientific and social improvement," as the members of the Augusta, Maine, association so aptly resolved in 1841.[2] Frequently the desire for mutual improvement led to the formation of institutes or schools where subjects as basic as writing and as complex as natural philosophy might be studied. This educational effort was at too low a level to be considered professional activity.

Temperance, morality, and cleanliness were advocated as means of raising the mechanic's lot, indicating a possible relationship with the reform movements abroad in society. Nor were economic issues ignored. The questions of fair compensation for skilled work done, of rapid payment of bills to mechanics by those able to do so, and of saving enough funds to retire and, as the *Mechanics' Mirror* put it in 1846, "be able to

---

1 "Mechanics Meeting," *Amer. Rr. Jour.*, January 24, 1835, p. 41.
2 "Mechanic Associations, Maine," *N.Y.S. Mechanic*, November 20, 1841, p. 5.

walk about with . . . staff in hand, an example and praise to those who do well" were all considered significant by the associations and periodicals.[3] There was, then, the expectation that middle-class standards would be adopted by all mechanics.

Although a concern was shown for hours of work, indicating that the example of factory regimen was beginning to extend to the artisan's shop, it was also made clear by members that such groups need not be "a union of journeymen mechanics for the purpose of forcing employers into the payment of such wages as the journeymen may demand."[4] If such groups were nascent forms of unionism or labor agitation, they were utopian indeed! In fact, the rejection of trade-unionism further enforces the image of entrepreneurial leadership.

The Albany-Troy, New York, group of mechanics' organizations and papers had their period of greatest activity in the 1840's, later than the Boston group which started in the 1820's and flourished in the mid-thirties, and differed from the older groups in several ways. The New York group was more politically oriented, had some tinges of anti-intellectualism, and generated a real and enduring controversy in mechanical occupations over the relative importance of theory and practice as guides to action and to subsequent success. The differences between the sophisticated Boston of the 1830's and the provincial capital of Albany in the 1840's do not alone explain the differences between the two groups.

While the Boston group seemed caught up in the fervor of reform and perfectionism endemic to New England in this period, the New York group had a more practical connection with political action, with pragmatic rather than transcendental thought. For example, the New York mechanics' organizations tried to end the prison labor system whereby convicted criminals learned and practiced craft trades of various kinds. The complaint against this system was cast in economic terms: that the prisoners' labor directly injured the livelihood of honest artisans in the outside world. This was the manifest function

---

[3] "To the Editor of the South Western Mechanic, Nashville, Tenn., from Robert Macfarlane," reprinted in "Editor's Table," *Mechanics' Mirror*, I (August, 1846), 197.
[4] *Ibid.*

of the system, but other latent functions may have subconsciously influenced the formation of these protest groups, such as the apparently unrecognized fact that the system undoubtedly had the effect of reducing the esteem or recognized social status which the public attached to such trades and mechanical occupations. A hint of recognition was present in the prison labor convention of 1841 in Albany. The delegates felt that not only was the economic competition unfair because of lower wages paid to the prisoners, but also that, when released, these artisans would hire themselves out as mechanics, poison the minds of apprentices and journeymen with whom they worked, and "call themselves mechanics, when really they are criminals."[5] The convention believed that the remedy was close at hand since "mechanics constitute $\frac{1}{5}$ or $\frac{1}{4}$ of the vote" and since the prisoners could easily be employed in stone quarrying, silk manufacture, or some other nonmechanical occupation.[6] Perhaps metal-working mechanics were hurt not so much *economically* by the prison labor system (elaborate and up-to-date machine shops were not likely the complement of many prisons) as socially by the threatened loss of public esteem which they felt would result from criminals engaging in their occupations.

Boston seems to have been the most active center for mechanics' mutual improvement associations. The *Young Mechanic* in 1832 listed seventeen separate societies for the improvement of the mind, at least half of which were associations of apprentices and journeymen-artisans.[7] The Boston Mechanics' Institute, founded in the mid-twenties by George W. Light, was representative of these.

Light, a globe-trotting, self-styled mechanic, was actually a mechanic in the machine-shop sense of the word. He had been an avid scientific experimenter in his youth and claimed to have experienced great embarrassment because of a membership rejection from a philosophical or scientific society. This experience gave him a burning desire to form mechanics' organizations for mutual improvement, thus furnishing a ladder for social

5 "Mechanics' State Convention," *N.Y.S. Mechanic*, December 11, 1841, p. 18.
6 "Transcript of a Mechanics' State Convention, Evening," *N.Y.S. Mechanic*, September 10, 1842, p. 127.
7 "List of Societies in Boston," *Young Mechanic*, I (June, 1832), 91.

mobility. Light devoted some years of his life to this kind of activity and became a journalist in the process.[8]

Light's own work experience was typical (even if his self-consciousness was not) and consisted of journeyman service in the New England metal- and wood-working shops which made machinery for cotton factories. Untypically, he studied mathematics in every spare moment, such as when returning by ship from a trip to Russia in 1823. Footloose in Boston in the mid-twenties, he drummed up support for his favorite idea, which bore fruit by the end of 1826 when the Boston Mechanics' Institute was formed. The mutual improvement idea led to lectures in "mechanical philosophy and chymistry [sic]" which drew such large public crowds that the original purpose of the organization seemed likely to be lost. At least it appeared so to zealots like Light, who editorialized in 1832 that "popular lecturing ought not to be the *regular* exercise of any institution, the professed object of which is *mutual* improvement."[9] Light had by this time become deeply involved in a series of journalistic endeavors, mostly unsuccessful financially, which were variously titled the *Boston Mechanic*, the *Boston Young Mechanic*, the *Young Mechanic*, and simply *Mechanic*.

In his mechanics' papers or magazines Light promoted the idea, taken up by most of the New England associations and pursued to an extent not so common in the New York group, that theory or first principles was more important to the mechanic than mere skill. As he expressed it editorially in 1832:

Of two individuals engaged in the same department . . . the one who is most familiar with the *principles* of his business will always have an advantage over him who is expert only in the use of his tools. Further than this, of two who are equally familiar with both the science and the art in question, he will have the advantage who has the most *general* information—the best disciplined mind—the firmest habits of mental industry, of vigilant observation of men, and things, of perservering effort, of thorough inquiry, and of systematic and accurate reasoning.[10]

George Light in his Boston magazine attempted to inspire

8 Article, presumably by George Light, "Memoir of a Mechanic," *Young Mechanic*, I (November, 1832), 163–67.
9 George W. Light, "Boston Mechanics' Lyceum," *Young Mechanic*, I (August, 1832), 125.
10 "To Young Mechanics," *Young Mechanic*, I (January, 1832), 1.

Machine tool shop at Richard Norris and Son Locomotive Works,
Philadelphia, Pa., 1855 (*United States Magazine of Science,
Art, Manufactures, Agriculture, Commerce and Trade*).

young mechanics with a vision of a world transformed by the steam engine into a paradise. Light contended that all the inventions and improvements of recent times paled "into insignificance, when compared with the extraordinary results which have followed the employment of steam as a mechanical agent."[11] He chided "our mechanics" for allowing the British to capture the railroad engine market by default. Light felt that it required "but the attention of some of our steam engineers to the subject" to produce superior American locomotives.[12] On a more practical level, Light ran a series of articles on the economy of heat and other scientific questions relating to steam power.[13]

Light felt that the defects of the mechanic in terms of theoretical knowledge could be remedied, but only by patient study. The New York mechanical editors also followed this theme on occasion, stating that "there is no royal road to science; nor can a man be a good mechanic by intuition."[14] Not only did scientific studies liberalize the mind but they could be brought to bear upon the solution of many otherwise difficult or impossible problems. One writer in Light's journal found a good example in Thomas Keyes, Jr., of West Boyleston, Massachusetts, who died in 1831 at the age of twenty-nine.

Keyes devoted himself night and day to mathematics, natural philosophy, and astronomy, particularly to mathematics, which he regarded as all-important. The author had often heard Keyes observe that mechanics were put at a disadvantage by their neglect of mathematics. Besides the practical advantage this study had to Keyes (presumably in the avoidance of error in the pursuit of a mechanical career), it instilled in his character the admirable trait of communicating information to others freely.[15] The author's orientation was clearly professional.

For one correspondent of the *New York State Mechanic* the acquisition of scientific knowledge and good personal habits

---

11 "The Steam Engine," *The Mechanic* (Boston), III (January, 1834), 21.
12 "Railways and Carriages," *Young Mechanic*, I (April, 1832), 54–55.
13 See *Young Mechanic*, I (1832), *passim*.
14 "Advice to Apprentices," *New York Mechanics' Magazine*, reprinted in *Boston Mechanic and Journal of the Useful Arts and Sciences* (hereafter cited as *Boston Mechanic*), IV (November, 1835), 214.
15 J. M. W., "Thomas Keyes, Jr.," *Young Mechanic*, I (April, 1832), 58–60.

were intertwined. "Fulton" advised mechanics to "so systema-
tize their habits and regulate their hours, as to make some
advancement in science every day."[16] Other writers also advised
a patient, step-by-step acquisition of knowledge that relegated
genius and inspiration to lesser roles.

Frequently the exhortation to science was placed on the practical level of applied science which could be useful in everyday
occupations. The mason could, by the use of a simple pulley
contrivance, lift heavy loads of bricks instead of carrying them
up a ladder; the carpenter, by using basic Euclidean geometry,
could be certain he was laying out a building at right angles;
the tanner could employ chemistry.[17] In one story a farmer is
astounded to learn that an artist-student, whom he has been
berating for impracticality, designed the scientific plough which
enables him to reap larger harvests.[18]

Occasionally, but not too often, editors and their correspondents developed a clear and invidious distinction between pure
and applied science, in favor of the latter. The *Mechanics'
Mirror* editorialized in 1846: "After this, experiment will test
truth of the dissertations of scientific men, and . . . no credence
will be given to any scientific work, unless it comes from the
workshop or studio of the practical mechanic."[19] In another
editorial, the same journal asked: "Shall mere scientific men,
and not operatives, be the high priests who shall explain the
laws and teach the principles which direct our respective mechanical callings?"[20]

This feeling at times drifted into a general distaste for the
traditional classical higher education. One editor noted "the
accumulation of good-for-nothing *classical* lore in our 'institutions.' "[21] Another correspondent in a report on Harvard College
wrote, "A mechanic may care little for a college; for the mechanic arts are not taught there, and the Greek and Latin he
does not need."[22] There was a call for practical handbooks and

---

16 "Advancement of Mechanic Interests," *N.Y.S. Mechanic*, February 12, 1842, p. 94.
17 "Address to Mechanics," *New York Mechanics' Magazine*, reprinted in *Boston
Mechanic*, IV (November, 1835), 200; "Economy," *Boston Mechanic*, IV (July, 1835), 121.
18 "Genius vs. Labor," *N.Y.S. Mechanic*, February 26, 1842, p. 112.
19 "Mechanical Philosophy," *Mechanics' Mirror*, I (April, 1846), 87.
20 "To the Mechanics of New York State," *Mechanics' Mirror*, I (January, 1846), 22.
21 "Mechanics Awake," *N.Y.S. Mechanic*, May 7, 1842, p. 189.
22 "Account of Harvard College," *Boston Mechanic*, IV (June, 1835), 78.

textbooks of the mechanic arts which would avoid the complexities of Newton's *Principia* and approach the principles of mechanics in a clear, simple fashion. Frequently the mechanic was absolved of blame for his lack of interest in science because of the "interminable volumes" full of "blind technicalities, quotations of foreign languages, and a dry uninteresting style" which he had to wade through.[23]

The awareness of a gap between theory and practice can be recognized in these early periodicals, and many editors and their correspondents expressed the hope that this gap could be reduced. "Marcellus" addressed the *Young Mechanic* and its readers in 1832 with this clear statement of the problem:

> The principal reasons which have been assigned for the slow progress of the useful arts, are drawn from the wide separation which has been made between science and art, and the fact that many of the greatest inventions have resulted from accident rather than superior knowledge. These considerations have tended to give to these pursuits the character of unintellectual employments. Whereas they require and call forth the best powers of the mind, and fail to receive the attention of the wise and talented only because they have been too much regarded as subjects of mere mechanical labor.[24]

Others thought such an indictment too strong; one addressing a mechanics' association in New York in 1842 described three classes of society: one which did little manual labor but to which society was nevertheless indebted, another which did nothing but manual labor, and the third, "a great class, who unite theoretical knowledge with practical skill"—the mechanics.[25]

It was clear, however, to the journalists and their correspondents that this one class with the ability to span the gap between theory and practice, between dilettante and day laborer, between science and art, was not receiving the social recognition to which it was entitled. Stressing science was one way of expressing concern over the status of the mechanic, and those who did so were interested in raising that status.

On a more practical level, mechanical journalists complained

23 "Introductory," *N.Y.S. Mechanic*, November 20, 1841, p. 5.
24 "Essay on the Mechanic Arts," *Young Mechanic*, I (March, 1832), 36.
25 "Mechanic Associations. Address," *N.Y.S. Mechanic*, January 15, 1842, p. 58.

that the respectable classes were not putting their boys out as apprentices at honest trades because of the onus attached to the name "mechanic." One writer noted that modern authors invariably showed mechanics in a bad light, "Always associated in their every movement and word with their trade, the every utterance groveling, and their appearance described as 'gaunt,' 'grim,' or 'dirty,' seldom allowing them the semblance of decent feeding, clothing or habitation, evidently placed in their pictures for the purpose of the favorable contrast it gives their heroes and 'better society' characters."[26] Another was bothered that the "wealthy and educated scarcely or never bind their sons to the trades." Why was this so, asked an anonymous correspondent of the *New York State Mechanic* in 1842: it was because they did not feel mechanical employment to be either as honorable or as profitable as other callings. "The reason is obvious—the association is not that of either refinement, education, or affluence. The mechanic has not either the political, moral, scientific, or religious influence, that the other pursuits possess."[27]

Certainly some of this ideological expression can be placed within the framework of the leveling influence of Jacksonian, egalitarian America, such as the statement in the *Boston Mechanic* in 1835 that "the old barriers are down. . . . Talent and worth are the only external grounds of distinction."[28] However, it is easy to carry this analogy too far and thus to miss the elements of a nascent professionalism in it. One of the editors of the *Boston Mechanic,* for example, pointed out to the readers that the mechanics were indeed on a low social and educational plane, but this resulted as much from lack of sharing of important practical secrets as it did from sheer lack of scientific knowledge of principles.[29] The word "profession" was used loosely by the mechanics' groups in referring to themselves, and they often indicated that the recognized professions might not be entitled to as much status and recognition as they had been receiving.

Attacking the position in society held by the lawyer had

26 "Condition of the Mechanics," *N.Y.S. Mechanic,* January 1, 1842, p. 47.
27 "Condition of the Mechanics," *N.Y.S. Mechanic,* January 8, 1842, p. 54.
28 "Respectability of Mechanics," *Boston Mechanic,* IV (November, 1835), 206.
29 "What is the Use of Learning," *Boston Mechanic,* IV (July, 1835), 113–15.

particular attraction, as he was a symbol of the nonproductive class of occupations. "Why should the merchant, the lawyer, the professor, the speculator and the many others less useful in society, arrogate to themselves superior claims to power, honors and profit," asked the *New York State Mechanic*.[30] Or, as the editors of the *Mechanic* put it: "There are hundreds of lawyers who would have made better mechanics . . . and . . . many mechanics who would stand high at the bar, had they been blessed with a liberal education."[31] Thomas Judd, addressing the Protection of Geneva Mechanics in 1846, put it even more strongly: "If you please, look at the legal profession. What ground do they occupy? It is a well known fact that, they compose but a very small portion of the community, and probably the most useless [but they] occupy more posts of honor and emolument than any other class."[32] For an answer, some proposed intellectual cultivation in the hopes that by study and sheer will the mechanic would be able to retake his stand "at the head of the professions, as your predecessors did of old."[33] Such was the object of the American Benevolent Education and Manual Labor Society, which was founded in New York City in 1845 and offered classes in modern foreign languages, agriculture, the mechanic arts, and bookkeeping, "so that our Mechanics may be better qualified to meet the professional man in our halls of legislation and in other stations of society."[34]

A few writers recognized the problem but did not see why there was no solution. Thomas Judd, after indicting the legal profession, noted their source of strength: "union, concert of action."[35] The order of business at a mechanics' convention in New York in 1842 was *"organization*, complete and entire organization." Yet, none of this literary activity bore any long-lasting fruit on a practical organizational level.[36] Why was it impossible for the mechanic to form a powerful professional

30 November 26, 1842, p. 5.

31 "Learning a Trade," *The Mechanic* (Boston), III (December, 1834), 354.

32 "Union among Mechanics," *Mechanics' Mirror*, I (October, 1846), 229.

33 "The Intellectual Cultivation of the Mechanic," *N.Y.S. Mechanic*, February 5, 1842, p. 28.

34 "Our Principles," *New York State Farmer and Mechanic*, April 10, 1845, p. 4.

35 "Union among Mechanics," p. 229.

36 "The Convention," *N.Y.S. Mechanic*, September 17, 1842, p. 137.

group which could assert considerable political and social influence in society?

For one thing, the conception of mechanic held by these groups was too diffuse and general to form a viable basis for professional activity. Since the membership represented almost every known trade and art, they could find no common knowledge base, an essential element of professionalism. In addition, this lack of a common knowledge base meant that united action was impossible except on a very few issues, such as abolishing prison labor competition. Only the common middle-class values of temperance, morality, sobriety, and industry were shared by the mechanics.

Secondly, the mechanics' organizations stressed that all occupations were equal, and their method of raising the status of the mechanic was to lower that of the older professionals, for example, lawyers. They had no clear conception of themselves as an elite, nor did their membership represent elites; this limited the possibility of professionalization among these groups. Elitist conceptions did influence the early professionalization of mechanical engineering in the United States, but the early mechanics' organizations had no part in that phenonemon.

There were perhaps dozens of possible professional specializations in these groups, if, and only if, a common set of problems and the self-consciousness of an elite developed. The breadth of the concept of the mechanic held by these groups was its greatest weakness in terms of the possibilities of professional culture. The concept included too much and excluded too little to become a viable basis for a mechanical profession, and it had no direct influence on the development of professional mechanical engineering. Not until 1880, when the American Society of Mechanical Engineers was formed, was there a group able to speak for the nation's mechanical engineers. One of the strongest factors influencing this ultimate specialization was the educational complex which grew so rapidly in the period after 1850. In many ways engineering education filled in the void left by the disappearance of the mechanics' organizations, but it brought a vigorous trend toward scientific and mathematical training and a sharper definition of occupational role.

ART **II**   INTERNAL DEVELOPMENT

"Unfortunately, that discrepancy between theory and practice, which in sound physical and mechanical science is a delusion, has a real existence in the minds of men; and that fallacy, though rejected by their judgments, continues to exert an influence over their acts."

W. J. M. Rankine
Scottish Engineer, 1855

There are two types of organizations within any professional group which exert formal influence on the nature and character of the profession. Educational institutions and professional associations, through their opportunities to determine the socialization process for the individual professional or would-be professional at different stages of his career, through their ability to set standards of knowledge and skill, through their self-assumed right to regulate titles and degrees, and in many other ways influence the internal development of the profession. There is, however, no guarantee that both institutions work for the same goals or even share the same basic concepts of what the profession ought to become. These may differ radically.

These institutions serve as vehicles through which one may explore the complex interrelationships and conflicts which seem endemic to an occupation in the process of trying to become a profession. In them one finds the highest aspirations of ideology confronted with the rudest facts of practice, and in them were sought solutions to the difference between the two. Of these institutions, the educational was undoubtedly the best organized and contributed most to what the profession of mechanical engineering ultimately became. The internal conflict in the profession over the nature of mechanical engineering education was both profound and enduring.

Many of the mechanics' institutes founded between 1820 and 1870 were actually attempts to combine the functions of the professional association and the school. The Franklin Institute, founded in 1824 in Philadelphia, had both a professional society and a technical high school. Speaking in 1828, Walter R. Johnson, principal of the school, underscored the practical nature of the courses given, including "the analysis and explanation of machinery." He hoped America could retain the wisdom, but abolish the fallacies of classical European higher education.[1]

---

[1] Walter R. Johnson, *Address Introductory to a Course of Lectures on Mechanics and Natural Philosophy, Delivered before the Franklin Institute, Philadelphia, November 19, 1828* (Boston: Hiram Tupper, 1829), pp. 3–15.

Similar institutes, lyceums, and libraries were founded in all parts of the industrial Northeast, and as early as 1828 the Ohio Mechanics Institute was established in Cincinnati. The original plan for the Massachusetts Institute of Technology was based to a large extent on this model, including a society of arts, a museum, and a school of industrial science and art, and emphasized the practical nature of the instruction to be given.[2] Various pressures caused most of these organizations to specialize in one of the three possible functions. Undoubtedly these institutes spread mechanical knowledge among mechanics and those who would later identify themselves as mechanical engineers.[3] However, they suffered from a diffuseness which characterized the conception of the mechanic, and only some of those which specialized in the function of technical education assumed the role of trainer of professionally oriented mechanical engineers.

In fact, the colleges and universities were among the first to recognize the possibilities of curriculum specialization and to develop a method for educating mechanical engineers when mechanical engineering emerged as an occupational specialty in the railroad, machine tool, and steam engine industries in the 1850's. Yale University created a professorship of industrial mechanics and physics in 1859.[4] Rensselaer Polytechnic Institute, which had become the major training school for American civil engineering, in 1862 inaugurated a course in mechanical engineering. Neither of these early attempts was immediately successful, and one died shortly thereafter. Yale did not begin to graduate a significant number of dynamical (an alternative term then in use) engineers until the early 1870's when the major trends in mechanical engineering education were already under way. Rensselaer's course seems to have existed on paper only, since it appears that no students ever took the course, no

2 Charles Barton Rogers, *Objects and Plan of an Institute of Technology* (2nd ed.; Boston: John Wilson and Son, 1861), pp. 5–28.

3 The popularity and usefulness of these institutes is attested to in a letter from George Escol Sellers to Coleman and John Sellers, September 27, 1850, in Peale-Sellers Papers. He notes that the Ohio Mechanics Institute "is crowded to suffocation during the evening and quite full all day."

4 Russel Henry Chittenden, "The Founding of the Sheffield Scientific School," *Inventors and Engineers of Old New Haven*, ed. Richard Shelton Kirby (New Haven, Conn.: New Haven Colony Historical Society, 1939), p. 100.

one ever graduated with the degree, and no professor was ever appointed in mechanical engineering; the course was dropped during a reorganization of curriculum in 1870.[5]

The Naval Academy did serve as a training institution for mechanical engineers. The curriculum, before the late 1860's, was basically a scientific one designed to turn out officers of the line who would know enough about physical science to be able to successfully command a modern steamship of war. Shortly after the end of the Civil War, the Academy added some teaching personnel, including the scientist Samuel P. Langley, and two young engineers who had gotten their experience aboard ship during the war. One of these men, Robert H. Thurston, helped raise the level of mechanical engineering instruction at the Academy, and the other, Erasmus Darwin Leavitt, soon left the Academy and began his own steam engine works, which served as a traditional shop training school for a host of influential mechanical engineers.

Thurston was born in 1839 in Providence, Rhode Island, of "old New England and English stock on both sides of the family." His father was a wealthy entrepreneur who formed the Providence Steam Engine Company in 1830 in partnership with John Babcock. Young Robert would probably have gone into his father's shop or the shop of one of his father's friends if it had not been for the influence of his schoolteachers, particularly Edward H. Magill (later president of Swarthmore College), who taught him science. Magill persuaded both father and son of the value of a college course in engineering and thus influenced the whole direction of the young man's life. Thurston intimated to friends in later years that he had at this early date (the mid-fifties) actually formulated the idea that engineering would shortly become a learned profession.[6]

Entering Brown University in 1856, Thurston pursued a broad, general education with special emphasis on science. His course led to a degree in civil engineering and a bachelor of

5 Palmer C. Ricketts, *History of Rensselaer Polytechnic Institute, 1824–1914* (New York: John Wiley & Sons, 1914), pp. 105–10.
6 William F. Durand, *Robert Henry Thurston* (New York: American Society of Mechanical Engineers, 1929), pp. 5–13.

philosophy (science) degree, after which he undertook the traditional apprenticeship in his father's firm. He failed to receive any preferential treatment and quit when he was offered a dollar a day to stay on with the firm. During 1860–61 he set himself up in Philadelphia as a "consulting mechanical engineer" and agent for his father's firm. This act of independence apparently earned him new respect at the elder Thurston's shops because he returned at full draftsman's wages in 1861. He immediately published his first technical article in the *Franklin Institute Journal* and applied for a commission as a third assistant engineer in the regular Engineer Corps of the United States Navy.[7]

Thurston's naval experience was valuable in that he met many mechanical engineers, was allowed to conduct some tests of equipment for the government, and finally got some actual shop experience (although a shop on board ship cannot really be compared to a machine shop on land). Thus Thurston's entire career experience, before he went to the Naval Academy as an instructor in natural philosophy, included only bits of actual contact with shop culture. But his experience brought him success since primarily it had been in occupational situations where his science education was appreciated. He rapidly became convinced that a thorough technical education was the first prerequisite to becoming a successful mechanical engineer. His lack of experience with shop culture no doubt contributed to his distrust of its efficacy as a training device for mechanical engineering.[8]

The would-be innovator was given little chance to implement his ideas by developing a mechanical engineering curriculum at the Academy. He did, however, upgrade many of the scientific courses and introduce new courses in such subjects as steam engineering. Thurston inaugurated a series of tests of lubricating oils, published articles, made technical studies of foreign steamships, and traveled to Europe in 1870 to study the Siemens process for making steel. These activities rapidly advanced his career. One major article in the *Franklin Institute*

[7] *Ibid.*, pp. 15–35.
[8] *Ibid.*, pp. 35–45.

Journal (on his trip to Great Britain) brought him such fame that it helped him to publish his way out of the Naval Academy by 1871.[9]

Thurston's theory of what engineering education should be was primarily a blend of French and German ideas. From France came an emphasis upon math and science and the concept of the high-level professional school; from Germany came the practice of setting up schools to train technical personnel at all levels, with the research institution at the top. Development in technical education in England had been slow and had contributed little to the ideas of engineering educators (though the English emphasis on pupilage was a model for shop culturists in the United States). Rarely, however, could engineering educators directly import European systems intact; more often the result of their labors was a blend of European ideas and American prejudices and attitudes toward education. These attitudes, relating to practical usefulness of education, can be traced from Jefferson's plan for the University of Virginia to an ever-increasing call during the first half of the nineteenth century for a system of colleges devoted to training in agriculture and the mechanical arts.[10]

The period from 1868 to 1872 was an active one for the inauguration of new programs in mechanical engineering. A major reason was the implementation of the Morrill Land-Grant Colleges Act of 1862. This act, pushed through Congress by a devoted proponent Senator Justin Morrill of Vermont, provided that the federal government turn over to the states certain lands, the proceeds of which were to finance the creation of agricultural and mechanical colleges. Many states did not take action regarding the act until the late sixties. When they did haltingly found these institutions, which were often based on the core of an older state or private college of some type, a

9 *Ibid.*, pp. 45–57.
10 An excellent survey of European forms of engineering education can be found in William E. Wickenden, *A Comparative Study of Engineering Education in the United States and in Europe* (Lancaster, Pa.: Society for the Promotion of Engineering Education, 1929). See also Charles R. Mann, *A Study of Engineering Education* (New York: Carnegie Foundation for the Advancement of Teaching [Bulletin No. 11], 1918) and Aubrey F. Burstall, *A History of Mechanical Engineering* (London: Faber and Faber, 1963), pp. 361–62.

natural step was to set up curricula in mechanics, industrial training, or dynamical or mechanical engineering.[11]

Typical of these mechanical engineering departments was the one established at the University of Wisconsin in 1870 under the stimulus of the land-grant funds. In fact, a whole range of programs in civil, mechanical, mining, and metallurgical engineering was inaugurated. University authorities appointed a professor of military science and civil and mechanical engineering, Colonel William J. L. Nicodemus. All the engineering programs struggled along during the next two decades, but the university administration took little interest in improving them. Mechanical engineering did not even have a shop. That situation prevailed until the late eighties and early nineties when interest in technical education blossomed again and with more permanent fruit.[12]

That the hastily conceived and implemented land-grant college engineering programs were unsuccessful in their first two decades is not hard to explain. The universities themselves were anxious to get land-grant money, but administrators frequently lacked faith in the value and future of technical education, particularly technical education in mechanical engineering. As a result the programs themselves were given little attention and many became a hodge-podge of lecture courses in basic mechanics, often taught by incompetent instructors.

This type of program was most common in the South, West, and Middle West, where the small demand for mechanical engineers served as an additional limiting factor. It held true, however, even in the Middle Atlantic states, where one might have expected a considerable demand for mechanical engineers with college training by 1870–85. Cornell University, which became the land-grant school for New York State, experienced severe difficulties with its mechanical engineering program, particularly from 1879 to 1885. Cornell's experience was so typical that it will be used later in the study as a test case in the development of mechanical engineering education.

11 Earle D. Ross, *Democracy's Colleges* (Ames, Iowa: Iowa State College Press, 1942); Edmund J. James, "The Origin of the Land Grant Act of 1862," University of Illinois, *Bulletin*, IV, No. 10.

12 Storm Bull, "Technical Education at the University of Wisconsin," *Wisconsin Engineer* (hereafter cited as *Wisc. Engr.*), III (January, 1899), 1–9.

In contrast to the inauspicious beginning of technical education in most colleges and universities was the experience of
Stevens Institute of Technology, founded at Hoboken, New
Jersey, in 1870. Edwin A. Stevens' will left $600,000 for the
purpose of founding an "institution of learning." The trustees
of the estate decided in 1870 that it should be an institute of
technology (since Stevens' father John Stevens had been a
gifted engineer who conducted some early experiments with
railroads) and appointed Dr. Henry Morton, a physicist and
secretary of the Franklin Institute, as its president. Morton
believed, and convinced the trustees, that the field of the institution should be narrowed to mechanical engineering.[13]

By 1871 Morton had induced Robert Thurston to leave the
Naval Academy and to become professor of engineering at the
new school. From 1871 to 1885 Thurston devoted his life to
creating a mechanical engineering curriculum that could serve
as a model for the pure technical school or professional school
with no admixture of general cultural education. The establishment of Stevens Institute of Technology, specifically devoted
to the training of mechanical engineers, added greatly to the
rationale for such programs at other universities, colleges, and
technical institutes.

Purdue University in Indiana was one of the first to follow
this model. Chartered in 1869 as a land-grant college, it received the needed push through a gift of money from John
Purdue. The school was not opened until 1874, and it bravely
announced courses in civil engineering, physics, and mechanical
engineering. Its president, Emerson White, openly rejected the
ideals of classical education and stressed the implementation
of the true meaning of the Morrill Act—practical education. He
feared that if the classical forces were allowed to adulterate the
technical curriculum, Purdue would become just another university and thus fail to fulfill its proper role as a mediator between science and industry. He carried this idea to extremes
by eliminating Latin and German, fraternities, and membership
in the State Oratorical Association.[14]

13 Durand, *Thurston*, pp. 61–62.
14 H. B. Knoll, *The Story of Purdue Engineering* (West Lafayette, Ind.: Purdue University Press, 1963), pp. 4–13.

White brought William F. M. Goss to the campus as professor of the department of practical mechanics in 1879. Goss's own training—operating a steam engine in his father's newspaper plant in Barnstable, Massachusetts—had been largely practical. After two years in the mechanical course at Massachusetts Institute of Technology (founded only one year before his entry), he came to Purdue at the age of twenty to head the new department of practical mechanics. The name of the department was compatible with Goss's capabilities and White's conceptions, and it simplified the problem of interesting manufacturers in the school. Goss grew with the situation and helped found a formal school of mechanical engineering in 1882.[15]

Mechanical engineering was given an additional faculty boost in 1883 when Lieutenant Albert W. Stahl, U.S.N., was assigned to the school. This assignment was part of a program inaugurated in 1879 by the Navy Department which helped many schools and colleges to raise the standards of their mechanical engineering faculty until 1896 when it was discontinued. The success or failure of many schools in mechanical engineering depended on the naval engineer assigned to them.[16]

A bill passed in February, 1879, allowed the President of the United States to detail an officer from the Naval Engineer Corps as a professor in any school, ostensibly to promote "knowledge of steam engineering and iron ship building." It was partly a way to employ naval engineers (the fleet had shrunk to a low level by the late seventies), and partly to provide a much-needed injection of superior teaching personnel into technical schools. These manifest functions of the program were complemented by the latent function of producing a complex and close relationship between naval engineering and mechanical engineering education.[17]

Between 1879 and 1896 forty-eight of these appointments were made to such diverse institutions as the University of Pennsylvania and the Chicago Manual Training School. All of the major land-grant universities had at least one such appointment. Many of these engineer-instructors developed and ex-

15 Ibid., pp. 160–62.
16 Ibid., p. 186.
17 Bennett, Steam Navy, pp. 732–33.

panded the mechanical engineering curriculum at their assigned schools and, having become teachers of engineering and lacking sufficient incentive to return to naval service (for reasons which will be investigated later), they resigned their commissions and took jobs as professors.[18]

One example was Assistant Engineer Jay M. Whitham, detailed to St. Johns College in Annapolis, Maryland, in 1883. In his role as professor of mechanical engineering, he inaugurated a course in *practical* geometry and, when the mechanical engineering program failed at St. Johns, Whitham resigned from the Navy at the end of his four-year tour to take a chair of engineering at the University of Arkansas.[19] Arthur T. Woods, assigned to the University of Illinois in 1883, created a mechanical engineering program with both theoretical and practical components, complete with laboratories and the newest equipment. Woods also resigned his commission to stay at Illinois and two years later took the Chair of Dynamical Engineering at Washington University in St. Louis.[20]

Some of these men became leaders in the field of mechanical engineering education, such as Mortimer Elwyn Cooley, who became dean of engineering at the University of Michigan and a president of the ASME,[21] and William F. Durand, who went from his first detail at the Worcester Free Institute (Massachusetts) to Michigan State College, Cornell University, and ultimately to a professorship of mechanical engineering at Stanford. He was also elected to the presidency of the ASME in 1925. The Naval Engineer Corps was thus a major contributor to the personnel of mechanical engineering faculties in the United States.[22]

The question of curriculum development leads to one of the necessary attributes of a profession: the possession of a systematic base of technical knowledge. Mechanical engineering

[18] *Ibid.*, pp. 733–36.

[19] *Ibid.*, p. 741.

[20] *Ibid.*, p. 740.

[21] American Society of Mechanical Engineers, *Diamond Jubilee Book of Facts* (New York: American Society of Mechanical Engineers, 1955), p. 31. For more material regarding Cooley, see *Scientific Blacksmith: The Autobiography of Mortimer E. Cooley* (New York: American Society of Mechanical Engineers, 1947).

[22] ASME, *Jubilee Book*, p. 33. For more material regarding Durand, see W. F. Durand, *Adventures in the Navy, in Education, Science, Engineering, and in War* (New York: American Society of Mechanical Engineers, 1953).

Machinist operating lathe at Richard Norris and Son Locomotive Works, Philadelphia, Pa., 1855 (*United States Magazine of Science, Art, Manufactures, Agriculture, Commerce and Trade*).

emerged in Europe before it did in the United States, and it was also in Europe that the theoretical basis was created, much of which was, of course, not specifically mechanical engineering, but basic physics. The basis for physics had been set by Newton's work in the seventeenth century; the nineteenth century saw new areas of interest open up, suggested in part by the practical problems of an industrial society.

Thermodynamics, or the science of heat and energy, was first explored by French and German engineers and scientists. Strength of materials was another subject of direct concern to the mechanical engineer, and here the basic work was done by Navier, a Frenchman. Elastic theory likewise was developed in France by Poisson. Experimental work in the stress on machine parts was done by Weisbach in Germany. A Scottish engineering professor, Rankine, studied fatigue in metals and the basic science involved in the workings of the steam engine. America had no scientist-engineers to rival these pioneers until late in the nineteenth century.[23]

American engineers recognized this debt to Europe for basic principles in their science. *Engineer* in 1860 complained that the United States had no first-class theorists in the field of mechanical engineering. America, so the editors felt, had plenty of second-class writers, however, and of these they listed Charles Haswell and Benjamin F. Isherwood of the Naval Engineer Corps and Holley and Colburn's work on railroad practice. The editors concluded: "Few scholars know anything of the practice of engineering and mechanics, and very few practical men can write for the press."[24]

Even the published works of the European scientist-engineers were not used extensively in America until the educational institutions forced them to the attention of a new generation of college-trained mechanical engineers. Thus, the large scale importation and dissemination of the basic works in the science of engineering was first undertaken by the technical schools

---

[23] Thorndike Saville, "Achievements in Engineering Education," *Centennial of Engineering*, ed. Lenox R. Lohr (Chicago: Centennial of Engineering, 1953), p. 213. N. T. Greene described this state of affairs retrospectively in a letter to Robert H. Thurston, December 2, 1893, in Robert H. Thurston Papers (hereafter cited as Thurston Papers), Cornell University Archives, Ithaca, N.Y.

[24] "Engineering Literature," *Engineer* (Phila.), September 27, 1860, p. 49.

through integrated curricula built around such books as Weisbach's *Mechanics* and Rankine's works on the steam engine.[25]

There was, however, one tool which was necessary if one was to read and understand these basic works. Higher mathematics—algebra, geometry, and, increasingly throughout the latter half of the nineteenth century, calculus—was required to cope with engineering literature. Rankine's *The Steam Engine and other Prime Movers* (1859), the first systematic synthesis of steam engine theory, demanded of the potential reader a relatively high mathematical sophistication. This is particularly significant since mathematics was, from the very first, the only area of engineering knowledge and skill which the technical school could claim as its special province.

Although shop culture maintained a high level of sophisticated experience in the actual building of machines and engines, mathematics, other than simple addition, was something that the shop could not or did not wish to take responsibility for in the training of mechanical engineers. As calculus and other areas of higher mathematics became increasingly indispensable to the practice of mechanical engineering, the educational complex gained in prestige. It was the one common denominator of all serious efforts in technical education in mechanical engineering. As *Technologist* noted in 1871: "In all institutions which afford a scientific preparation for the different departments of engineering, the differential and integral calculus appears as one of the studies of the course." In addition to its indispensability as a working tool, the editors believed that higher mathematics led to a clarity of mind and development of logical facilities which would aid a young man in every endeavor, especially in business life.[26] *American Engineer's* editors thought in 1881 that higher mathematics was indispensable to the practical engineer. Those who thought otherwise would, in the editors' opinion, set up a "great barrier to progress."[27]

25 Ralph R. Shaw, *Engineering Books Available in America Prior to 1830* (New York: New York Public Library, 1933), p. 18, is helpful on this subject. English books predominated with 475 titles, French books came next with 188, and there were only 29 in other languages.

26 "The Value of the Calculus as a Study," *Technologist: Especially Devoted to Engineering, Manufacturing and Building* (hereafter cited as *Technologist*), II (July, 1871), 183.

27 "Higher Mathematics in Practical Engineering," *American Engineer* (Chicago) (hereafter cited as *Amer. Engr.* [Chicago]), II (March, 1881), 43.

There was, however, no universal feeling that the necessity for mathematics was a good thing for mechanical engineering. A controversy raging in the technical magazines in 1897, and particularly in the *American Machinist*, illustrates the conflicting views. Frederick Remsen Hutton, secretary of the ASME, began the exchange in March with an attack on the tendency of the mechanical engineering education field "to make proficiency in *mathematics* the criterion of capacity for professional engineering work to the prejudice of other sciences." Some men, Hutton went on, graduated from the technical schools with a talent for math and nothing else.[28] In April Charles H. Benjamin of Case School replied to this attack by claiming the role of the technical school to be "to teach students the things they cannot learn outside."[29] In May Edgar Kidwell (a professor of mechanical engineering) and F. W. McNair (a professor of math) did Benjamin one better. They maintained in a letter to the editors that "the craftsman's experience, dexterity, and mechanical intuition cannot supplant skill in mathematical analysis."[30] The series ended rather undramatically with two letters by technical graduates extolling the vital, indispensable quality of higher mathematics in engineering education.[31] Hutton, though numerically outnumbered, did not represent the old mechanic resisting a new tool he did not understand. He was a technical graduate of the Columbia University School of Mines and also had an undergraduate degree. He became dean of the engineering faculty at Columbia, authored many textbooks on the mechanical engineering of power plants, and acted as an engineering consultant, in addition to serving twenty-four years as secretary of the ASME and filling its presidency in 1907.[32]

Another role which educational institutions attempted to fulfill was the direction of research projects in engineering. The

[28] Letter to the Editor, *Amer. Mach.*, March 18, 1897, pp. 218–19.
[29] Letter to the Editor, *Amer. Mach.*, April 29, 1897, p. 328.
[30] Letter to the Editor, *Amer. Mach.*, May 6, 1897, pp. 346–47.
[31] Charles L. Griffin, Letter to the Editor, *Amer. Mach.*, May 27, 1897, p. 400; E. S. G., Letter to the Editor, *Amer. Mach.*, June 10, 1897, p. 437.
[32] ASME, *Jubilee Book*, p. 37; Frederick Remsen Hutton, *A History of the American Society of Mechanical Engineers from 1880 to 1915* (New York: American Society of Mechanical Engineers, 1915), pp. 117–18. Hutton's ancestors had come to Kingston, N.Y., in 1642, and his father was a prominent pastor in New York. His early training had been in private schools. Frank A. Taylor in *DAB s.v.* "Hutton, Frederick Remsen."

profession as a whole agreed by the 1870's that engineering research laboratories ought to be established to investigate such questions as the causes of steam boiler explosions, stresses in machine tools, strength of materials, and wearing characteristics of steel and iron rails—in short, everything from the most scientific to the most practical questions.[33] As no other bodies took action, engineering schools quickly latched onto research as a major and legitimate concern. Robert Thurston petitioned the trustees of Stevens in 1874 for the right and the means to establish a laboratory for technical research or testing laboratory to investigate (with funds derived primarily from interested railroad executives) resistance of trains, efficiency of locomotives, strength and other characteristics of materials, and the value of fuels and lubricating materials.[34] It was set up and was moderately successful. The programs for locomotive testing set up later at Purdue and at Ohio State University were much more successful and gave technical education a real foothold in organized research.

It is a mistake, however, to think of technical education as a monolithic enterprise with common goals, united against ignorance, trying to train only professional engineers. From the very first, institutions teaching mechanical engineering or the mechanic arts were differentiated into types, each trying to fulfill a different role in education and in engineering. One type was the institution which set out to train scientifically competent, professional mechanical engineers. This type was further divided into the engineering school which was part of a university or college, with engineering as simply one of the majors for undergraduates, and the strictly technical school or professional school where only engineering was taught. The latter could be completely independent (Stevens Institute and Case in Cleveland) or it could be part of a large university (Sibley College of Engineering at Cornell University). These schools usually put great stress on higher mathematics and research and sometimes on general science courses, such as

---

[33] One engineer suggested that some capitalist who wished to benefit mankind should, instead of founding a library or art gallery, found "a laboratory of engineering research." W. K. Stevens, "Engineering Ignorance," *Amer. Mach.*, August 7, 1890, p. 6.

[34] Durand, *Thurston*, pp. 70–72.

physics. Frequently they also had workshops where practical knowledge could be gained, but this aspect was a source of conflict among educators and practitioners of mechanical engineering alike. These schools developed admission standards which put a premium on graduation from high school by requiring knowledge which could be most easily gained there.

The second type of educational institution was the trade or industrial training school, which had as its object the training of highly skilled mechanics who could step into supervisory positions in industry. In contrast to the mechanical engineering schools, these institutions stressed practical shop training rather than basic theory. One example of this type was the Worcester Free Institute in Worcester, Massachusetts, where great emphasis was laid on actual production of items for sale by the students in the shops. It was frequently believed that training for foremanship in a shop or plant would ultimately lead the gifted into mechanical engineering. Rather than being concerned with the education of professionals, these trade schools were devoted to training which would enable the student to immediately get a job as a foreman, superintendent, or even machinist. Their admission standards did not preclude those without high school background.

A third response to the demand for technical education was the movement to create manual training schools at the high school level, often as a division of a large public high school. The major purpose of these institutions was to train workmen, mechanics, and machinists for industry. Professional mechanical engineers were little interested in this third type of training as a part of education for the profession. They showed great interest, however, in the development of, and relationship between, the trade school and the engineering school.

By the 1890's the education of engineers was becoming a profession in itself. Twenty-five mechanical engineering teachers met in June, 1891, at Ohio State University to form an engineering education association.[35] This group was superseded by the formation in 1893 (during the World's Columbian Ex-

---

35 A. J. Wiechardt, Letter to the Editor, "A Proposed Engineering Society," *Amer. Mach.*, June 11, 1891, p. 5.

position in Chicago) of the Society for the Promotion of Engineering Education, composed of men teaching civil, mechanical, mining, and electrical engineering. This powerful group was able in a few short years to gain complete control of curriculum, admission standards, and other basic constituents of engineering education. Although it maintained close relationships with the four major engineering societies, the organization's membership and leadership were composed of engineering educators, not practicing engineers.[36]

A major problem to be solved by educators was the potential demand for mechanical engineers, foremen, and skilled workmen. If technical education was to continue to increase at a geometrical rate (and it was: the number of institutions offering formal engineering programs increased from 17 in 1870 to 110 in 1890) some effort had to be made to determine whether there would be jobs for the graduates at the proper level (consistent with their education). The supply of mechanical engineers had to be kept equal to the demand and not outrun it if a profession was to be created and maintained.[37]

As a matter of official ideology, the engineering educators themselves asserted that the demand for trained mechanical engineers far exceeded the supply. In letters to the editors of mechanical engineering periodicals they frequently cited letters which they had received from manufacturers offering good positions to young technical graduates. Some technical editors, always more down to earth than engineering educators, also expressed the belief, as did *Technologist* in 1870, that the most pressing need of the country was for "a large increase in the number of engineers, architects, and educated superintendents of technological processes."[38] Such sentiments were echoed by journals throughout the country.[39]

The first decade of the twentieth century produced enough

[36] Frederick T. Mavis, "History of Engineering Education," *Centennial*, ed. Lohr, pp. 193–94.

[37] B. E. Smith, W. C. Weeks, and R. J. S. Carter, "Opportunities for Young Technical Graduates in the United States, An Accurate and Timely Discussion by Experienced Authors Residing in the East, West and South," Engineers' Society, University of Minnesota, *Year Book* (hereafter cited as Minn. Engr. *Year Book*), XVI (April, 1908), 33–37.

[38] "Technological Education," *Technologist*, I (February, 1870), 1.

[39] See particularly "Advantages of a Technical Education," *Practical Mechanic*, I (December, 1887), 8.

demand for mechanical engineers to justify a University of Wisconsin professor's statement (in 1909) that for the past ten or twelve years there had been "several times as many positions open as there were graduates."[40] In 1905 Dexter S. Kimball of Cornell affirmed the great demand with statistics which showed that Sibley College graduates were mostly in either mechanical engineering itself or related engineering and business fields. Very few had been forced by circumstances to seek work for which their training had no application.[41] At Wisconsin, manufacturers apparently did not write on any extensive scale to offer jobs until about 1900.[42] At Cornell, however, there were examples of the practice as early as 1880, when T. H. Fuller of the Globe Nail Company in Boston wrote the president of the university asking for first-class mechanical engineers to be trained as heads of special departments.[43] Even the generally optimistic *American Machinist* did not believe in 1906 that employers crowd "around the doors of college auditoriums and struggle for a chance to collar young graduates and carry them off to work at high salaries."[44] In another editorial the editors demanded more industrial and trade schools, but fewer schools of mechanical engineering. The editors frequently expressed the belief that many technical graduates were forced by lack of demand for their services to take menial jobs at low salaries. What the country needed was fewer mechanical engineer chiefs and more foreman-machinist Indians.[45]

Adding to such fears was the repeated circulation of rumors about oversupply of technical graduates in Germany from the 1880's to 1910.[46] One rumor indicated that all German university-trained engineers were considering organizing to demand minimum living wages, a move which spelled death for profes-

---

40 Professor Shepardson, response to a talk by H. J. Gile of Minneapolis General Electric Company, *Minnesota Engineer* (hereafter cited as *Minn. Engr.*), XVII (January, 1909), 86.

41 "Overcrowding in Technical Schools," *Amer. Mach.*, November 30, 1905, pp. 753–54.

42 F. E. Turneaure, "The Engineering School in the United States," *Wisc. Engr.*, IX (June, 1905), 274–75.

43 March 8, 1880, in Sibley College Papers, Cornell University Archives, Ithaca, N.Y. (hereafter cited as Sibley College Papers).

44 "The Demand for Graduates," *Amer. Mach.*, July 26, 1906, p. 127.

45 "Technical Education," *Amer. Mach.*, March 29, 1900, p. 283.

46 See, for example, "Technical Schools," *Amer. Mach.*, February 27, 1886, p. 8.

sionalism.[47] Referring to the German engineers' plight, *American Machinist* noted in 1905 that there were 40,000 practicing engineers and 15,000 engineering students in America.[48] Such a growth ratio of new practitioners could lead to serious overcrowding in the profession. Actually, both sides were correct in their way. By the turn of the twentieth century the number of available candidates for the top jobs in the innovative machine and engine shops was far greater than could ever be absorbed by the shops. Simultaneously, however, new jobs and roles were opening up for individuals with training as mechanical engineers, which absorbed the excess output of the engineering schools. How one viewed the job market in 1900 depended solely upon his own expectations and ability.

This raises the question of the level of ability actually exhibited by the technical graduates from mechanical engineering schools. Most commentators agreed that the technical graduate generally had a higher level of mathematical sophistication and skill than the uneducated mechanic in the shop, but there was sharp disagreement on whether he was able profitably to apply what he knew to practical problems in the shop. The mechanics' magazines of the 1870 to 1910 period were full of folksy stories about the smart-aleck young technical graduate who spouted mathematical terms and berated practical mechanics for their lack of theoretical knowledge, but who could not himself perform the simplest of mechanical jobs. Frequently, but not always, it was implied that in addition to being ignorant of practice the college-graduated engineer had his mathematics and theory garbled. Always he was rescued (just before blowing up an engine or ruining a machine) by a grisly, greasy old mechanic who solved the problem with the direct application of a basic scientific law which he had, of course, learned the hard way—by experience.[49] Thus, it was not a case of whether science was useful—it was instead a question of whether useful and correct science was the province of the experienced shop

---

[47] "Over-Production of Engineers," *Morning Ledger* (London), reprinted in *Amer. Mach.*, December 22, 1904, p. 1704.

[48] "The Plight of Technical Graduates," *Amer. Mach.*, July 20, 1905, p. 95.

[49] For examples, see Dixie, Letter to the Editor, "Is the Practical Man in Danger?" *Amer. Mach.*, August 4, 1904, p. 1031.

man or of the young college-trained mechanical engineer. Science was not being questioned; merely the proper method of gaining a useful knowledge of its basic principles was being debated.

Yet much of the ideological expression of dissatisfaction with technical education and its products sounded like the last growls of old, practical mechanics about to be displaced by college-trained men proficient in skills of which the former were ignorant. A noncollege-trained New Mexico mining engineer wrote to the *Scientific Machinist* in 1896 and asked if everyone would have to become a scientific man in order to cope with the hordes of young technical graduates descending upon the industrial world. If so, the engineer elected to stay in his mining plant until "improvements reach me and crowd me out."[50] A series of articles in the *American Machinist* in 1904 explored the question "Is the Practical Man in Danger?" The series prompted letters from practical men who labeled mechanical engineering graduates conceited and brainless and suggested that the only way they could take over the jobs of practical men was if the practical men lost their practicality and the technical graduates developed some.[51]

The idea of the old practical mechanic resisting progress, theoretical knowledge, and the scientific way of doing things was a simple explanation, and perhaps to many an appealing one. It was partly true. *Engineer* noted in 1860 the hundreds of men in authority (both nonengineering administrators and mechanics) who were resisting reform and their "distress" at their hopeless inability to cope with complex engineering situations.[52] *Mechanic* reprinted an article on "the manufacturers who have risen from the bench, without acquaintance with technical science, [who] constantly feel themselves at a disadvantage."[53] *Wisconsin Engineer* published a letter from "The Young Engi-

50 A. W. G., letter "From a New Mexico Cañon," *Scientific Machinist*, April 15, 1896, p. 3.
51 See particularly Corneil Ridderhof, "Is the Practical Man in Danger?" *Amer. Mach.*, June 16, 1904, pp. 809–10. By 1916, however, a study showed that out of 120 representative firms, 60 preferred engineering graduates to shop-trained men, 40 preferred the shop men, and 20 had no preference. Mann, *Engineering Education*, p. 20.
52 "Railroad Men," *Engineer* (Phila.), August 30, 1860, p. 17.
53 "Science in the Workshop," *Trade Review*, reprinted in *The Mechanic* (New Jersey), XII (November, 1882), 1.

neer," who felt that employers divided clearly into categories of college-bred and self-made, the former crediting, the latter discounting, formal mechanical engineering training not backed by experience.[54] The image has indeed crept into modern accounts of the history of engineering, which frequently state that the American engineering school had to fight prejudice against their product on the part of "old-time practical engineers." This ideological picture of progressive and scientific engineering educators fighting old fogyism and backwardness does not, however, fit the facts. The conflict was more than a battle between two systems of education; it was a deeper struggle between two cultures—school and shop—for control of the whole process of socialization, education, and professionalization of mechanical engineers in America.

[54] V (May, 1901), 210.

CHAPTER 4   SCHOOL VERSUS SHOP:
CONFLICT AND COMPROMISE

As mechanical engineering education became a field in itself, related to, but separate from, engineering practice, the educators discovered that to make their existence necessary they had to discredit shop culture as an educational institution. This involved two modes of action: one, to point out the deficiencies of shop culture as an institution for selecting, educating, socializing, and professionalizing mechanical engineers; and two, to formulate an alternative culture in technical education which could supply both manifest and latent functions of shop culture, making it unnecessary. Because of structural changes in the role of the mechanical engineer in industry, they were able to succeed in this undertaking.

The attack on shop culture took several forms. One was to stress its stagnation and lack of intellectual depth. Professor C. H. Benjamin, a leader of the school forces, wrote to the *American Machinist* in 1899, saying that technical graduates were almost invariably better than shop boys. He described the "narrowness and shallowness of the boy who has grown up with the shop without further education. Naturally, the man at the head of the shop thinks that boy all right, since the latter is but a reflection or miniature copy of himself."[1]

Others, such as Professor S. W. Robinson of Ohio State University, attacked the whole idea that the word "shop" had any possible connection with education. Opting for the term "mechanical laboratory" for his shop courses, he asked, "Why not call a chemical laboratory a medicine shop . . . likewise, why call a mechanical laboratory a shop?" Shops were, he continued, for making articles for monetary compensation—

1 "Some Testimony as to the Value of Technical Education," *Amer. Mach.*, March 2, 1889, pp. 159–60. In contrast, note the comments of Coleman Sellers to a friend regarding the superiority of American over British workmen, whom he described as downtrodden and too specialized. He wrote Henry Greenwood, when employed at William Sellers and Co., that "I could take you into our establishment and you could there see three hundred workmen all capable of conversing on general subjects with intelligence." Sellers to Greenwood, April 18, 1863, in Peale-Sellers Papers.

laboratories were for education.[2] Robinson was one of the prime movers behind the establishment of the Mechanical Engineering Teachers Association in 1892. At its first annual meeting in Columbus, Ohio, he gave a paper on the role of shop work in mechanical engineering education. In it he stressed that extensive shop work in college amounted to little more than apprenticeship, or the trade school, where the main object was to train a skilled workman, not a mechanical engineer. He concluded: "We have no use for 'shop work' in a school graduating men as mechanical engineers."[3]

A paper before the ASME on the Worcester Free Institute elicited the comment from William Kent, who had been Thurston's assistant at Stevens, that because of its heavy emphasis on shop work and recreating the shop environment Worcester was not a school of mechanical engineering at all. Such a reconstruction of shop culture would produce nothing but mechanics and machinists who could make two dollars a day (though he felt that this was a very commendable role for Worcester).[4] This was no attack upon the commercial aspect of shop culture since all mechanical engineers had to face commercial realities in their work.

Educators applauded the study made by ASME president James Mapes Dodge in 1903, which determined statistically the money value of technical training. Dodge took a group of forty-eight individuals, twenty-four of whom had had some technical training, twenty-four with only shop experience. He was able to show with graphs that the technically trained man made a larger return on his investment in education in terms of life income. The shop man started higher, but was soon left behind.[5]

In formulating their answer to the problem of training mechanical engineers, educators placed some emphasis on the idea that to know a subject was not to be able to teach it. An

---

2 "Forgings from Sibley College," *Amer. Mach.*, March 9, 1893, pp. 5–6.
3 Reported in "The Mechanical Engineers Teachers Association," *Amer. Mach.*, August 4, 1892, p. 9.
4 Comment on George I. Alden, "Technical Training at the Worcester Free Institute," ASME, *Trans.*, VI (1884–85), 528–29. See Carl W. Mitman in *DAB s.v.* "Kent, William."
5 "The Money Value of Technical Training," ASME, *Trans.*, XXV (1903–4), 40–48.

editorial on technical education in 1872 stressed that the trouble with the old apprentice-shop system of education was "that it entirely ignored the superior value of a good teacher, and was based on the false theory that a good mechanic, or a good doctor, or a good lawyer, must of necessity be a good teacher, while we know that the reverse is too frequently the case. Teaching is a distinct art."[6]

Technical education could offer potential mechanical engineers professional instructors, who provided students with abstract knowledge of the scientific principles of engineering and also (through mechanical engineering laboratories) made it clear how these principles could be applied in practice. This new education, as the dean of Wisconsin's engineering school described it in 1902, "cultivates the useful and neglects the useless," was "humble and democratic," and was a "powerful tool, as a means to an end, as an instrument of pleasure and profit to oneself and of service to the world." The dean posed this "new education" in contradistinction to classical education, not to shop culture, but the image of technical education as both eminently practical and scientific was a common and powerful one.[7]

Robert Thurston saw an even grander plan, with "trade schools in every town, technical schools in every city, colleges of science and the arts in every state, and a great technical university as the center of the whole system."[8] Such a system was taking shape by the 1890's, and by 1902 an engineering professor at the University of Illinois could state with some truthfulness that technical education could now be regarded as a prerequisite to the "proper practice of the engineering profession."[9]

There was, however, a group of men who were critical of the broad claims of technical education and who rose to the defense of shop culture. One finds in them a pattern of character-

6 "Our Technical Schools," *Industrial Monthly*, I (March, 1872), 85.
7 "The New Education," *Wisc. Engr.*, VI (April, 1902), 205–6.
8 "Our Progress in Mechanical Engineering," ASME, *Trans.*, I (1880), 452–53.
9 Alfred Hume, "Some Thoughts on Engineering Education," Engineering Association of the South, *Transactions* (hereafter cited as Engr. Assoc. South, *Trans.*), XIII (1902), 91.

Machine shop of the Harris-Corliss Engine Works, Providence,
R. I., 1879 (*Scientific American*).

istics which is surprising only if one accepts the ideological concept of shop culture formulated by the education professionals. The defense of shop culture was not a cry from old-time mechanics about to be replaced by smart-aleck young technical graduates; it was instead a reasoned defense of a viable, working institution threatened by changes in the industrial world. This institution, the defenders felt, could do a better job than the technical school and could not be dispensed with except at a great loss to the profession. They doubted the ability of the technical school to become the functional equivalent of the older system of shop culture.

One part of the case made for shop culture was that the commercial conditions in which the shop operated produced a more healthy respect for accuracy than did the school situation. As one engineer expressed it in a meeting of the ASME, the trouble was that the school required only 75 per cent accuracy and "the profession of engineering is, above all others, a hundred per cent profession."[10] Highly successful engineers, college graduates and nongraduates, frequently attacked the engineering schools for their low standards, not of admission, but of achievement.

There was also a mystical emotional experience about the shop, according to many leading engineers. Frederick Winslow Taylor on many public occasions suggested that young graduates had to learn how to make a living, to learn the viewpoints and methods of thought of the workmen, how to talk with them on their own level, to have respect and kind regard for them. One learned things better, Taylor believed, from a few knocks on the head than from an engineering book. Taylor himself gave up the chance to go to college in order to take the traditional shop apprentice route to mechanical engineering. He got his technical degree at Stevens later; nevertheless, he once said that one year in a machine shop was worth twenty years in school, and he wrote Stevens Institute's President Alexander C. Humphries that the prominent and successful

---

[10] James Mapes Dodge, comment on Calvin M. Woodward, "The Training of a Dynamic Engineer in Washington University, St. Louis," ASME, *Trans.*, VII (1885–86), 773.

manufacturers he had talked to believed there were no technical graduates in responsible positions who had not once been workmen. There were two reasons why engineering students did not take shop jobs after graduation, Taylor wrote: the low pay and a feeling that manual labor was hard and irksome "after they look upon themselves as engineers." The resulting lack of experience drove college graduates into the drafting room, which Taylor felt was the equivalent of the bookkeeping room in a mercantile establishment. One year of practical work was a necessity for the young engineer. Taylor believed that technical education was important but not so much as shop culture. Ironically, he did as much as anyone to help destroy it.[11]

Charles E. Emery, an eminent consulting steam engineer, advised the 1880 graduating class at Stevens to value their education, but to be humble and willing to learn from old hands in the shop.[12] Besides, as the *American Machinist* pointed out in 1886, "hereafter more book knowledge of technical matters will not be of much value, and this will be due to the fact that shopmen generally will acquire this knowledge."[13] William O. Webber, applauding the success of Worcester Institute, wrote the *American Machinist* that he preferred men of experience to theoreticians and that men must enter engineering from machine-shop experience. He disagreed sharply with some professors who felt "that the three or four years of shop life unfit the boy for study." Instead the shop offered a perfect opportunity to learn rights and duties.[14]

Prevalent also was the fear, as expressed by J. S. Walker, president of the Engineering Association of the South in 1898, that "When we come to educate the rising generation of engi-

11 Comment on John Price Jackson, "College and Apprentice Training," Engr. Assoc. South, *Trans.*, XXIX (1907), 499; Taylor to Humphries, March 30, 1908, in Taylor Collection.

12 "Address to Student Engineers," *Amer. Mach.*, January 10, 1880, pp. 4–6. Emery's ancestors landed at Boston in 1635. He was educated at Canandaigua Academy, where he had a reputation as a scholar. In 1863 he was married to the great-great-granddaughter of General William Livingston, colonial governor of New Jersey. See Edna Yost in *DAB s.v.* "Emery, Charles Edward."

13 "Education for Mechanics and Engineers," *Amer. Mach.*, July 3, 1881, p. 8.

14 "Admission Requirements in Engineering Schools—the Apprentice System," *Amer. Mach.*, March 12, 1896, pp. 289–90.

neers, . . . we may have them too much educated and refined for the work they are to do." Walker pointed out that only in the shop could a man acquire the workman's knowledge of materials, tools, and plant and the "foreman's ability to control and manage the working force."[15] This was no greasy mechanic speaking, but one who felt gentility's place was in the drawing room, not in the shop. Calvin W. Rice, second permanent secretary of the ASME from 1907 to 1934 and a technical graduate himself, remembered his days as an apprentice with respect, not with horror. He cautioned against removing the conditioning agent of the shop from the educational process of every young mechanical engineer.[16]

The matter of selection of the proper persons to become mechanical engineers was another source of conflict. William F. Durfee told the ASME in 1885 that only the shop could perform this job. The schools, he said, used criteria which selected those least fit for engineering, those lacking inborn "intuitive practical sense of the fitness of things." The schools, unlike the shops, were turning out inexperienced, useless products, "clad in sheepskin, 'cum maxima laude', and decorated with some such titular abomination as 'dynamical engineer.' "[17] Durfee was a college-educated engineer, did some of the first truly scientific metallurgical work in an industrial plant in the United States, and owned a large engineering library.[18]

The foundation of mechanical engineering training, according to Frederick A. Halsey, must be the machine shop, not the high school (which was rapidly becoming the only way into the engineering schools). He preferred the shop, among other reasons, because it was more in the tradition of democratic America, where every man has a chance. He agreed with Durfee that the ever-rising admission standards in the engineering schools were cutting out those most likely to succeed in me-

---

[15] "Annual Address of President," Engr. Assoc. South, *Trans.*, IX (1898), 70–72.

[16] Comment on Jackson, "College and Apprentice Training," p. 519. In taking this line the American shop elite had a good model in the British shop world, where university training was considered unimportant and it was thought that the mechanical engineer's training should be primarily in the workshop. Burstall, *Mechanical Engineering*, p. 287. This did not characterize the continental European countries. Wickenden, *Comparative Study*, pp. 7ff.

[17] Comment on Woodward, "Training of a Dynamic Engineer," p. 777.

[18] Hutton, *History*, pp. 136, 142.

chanical engineering, those with shop experience.[19] Halsey was a Cornell graduate himself, a leading scientific management expert, and part of a student movement (when at Cornell) to oust an incompetent professor and *raise* academic standards. This pattern was indicative; those who opposed high standards of admission to the schools generally deplored low standards of excellence in them.

The leading critics of technical education, the defenders of shop culture, were members of the mechanical engineering elite. They were not opposed to technical education but believed that shop training was in many respects superior and should definitely precede formal education. They had generally received a technical college education themselves (though frequently after their shop experience), and they were highly scientific men, often innovators in the application of science to their particular fields, such as Fred Taylor, Fred Halsey, and W. F. Durfee. In short, they were representative of the cream of mechanical engineering practitioners.

Even though they frequently couched their opposition to the exclusiveness of technical schools in terms of democratic ideology, one can infer that they feared a loss of the concept of an elite in mechanical engineering. The rapid expansion of technical education, particularly in the Midwestern land-grant colleges, supplied thousands of mechanical engineers where previously there had been hundreds. Furthermore, these new engineers tended to come from a lower socio-economic level. Increasingly their background was the high school, not the shop, and they often had a disdain for the dirt and roughness of the shop, frequently finding it difficult to adjust to their initial roles as apprentices. The mechanical engineering elite came, as a whole, from a higher social and economic class (often the upper class) and resented the destruction of the institutions they had built around the shop. They did not fear the shop; on the contrary, they were convinced that any man who rose from the shop into mechanical engineering deserved to be there. They doubted, however, the claim of many of the young men who bore the title "M.E.," bestowed by a technical

19 "Discussion on Paper No. 850," ASME, *Trans.*, XXI (1899–1900), 733–35.

institution, to mechanical engineering status, unless their training had been supplemented by the traditional and prior shop experience.

Some of this elite were entrepreneurs, to be sure, but many were "pure" engineers with only marginal financial interest in the companies for which they worked or acted as consultants. Their opponents, the educators, were definitely bureaucrats, and the conflict between them over the nature of technical training and its value can be cast in terms of entrepreneurial versus bureaucratic ideas and concepts. The educators thought of training the average student by means of orderly institutional processes.[20] The elite emphasized the ability of shop culture to uncover the natural genius and to raise him above other men. Beyond all else the elite opposed the concept of averagism in education. The engineer they idealized was no bureaucrat; he was an entrepreneur sometimes, but always a leader and innovator.

Titles and words were symbols in this conflict. The term "shop" was frequently replaced by professional educators with terms such as "laboratory" in descriptions of practical machine experience offered in the schools. Likewise, the use of the term "dynamical" instead of "mechanical" to describe the branch of engineering was favored by the educators probably because it avoided the root "mechanic" and gave mechanical engineering greater intellectual status. The elite successfully opposed this change in name and usually found it, as did Durfee, a "titular abomination." The educators did exhibit more concern for the achievement of professional status by mechanical engineering, in part because this was the only way that many of their graduates would be able to avoid becoming wage-earning workmen. The elite, with status already assured—frequently by social class, sometimes as entrepreneurs, and by mutual recognition within their own world of mechanical engineering—seldom felt the need for advanced professionalizing activity. The difference was in the relative security of status felt by each group for itself

---

[20] In a letter to Frederick W. Taylor, April 3, 1908, in Taylor Collection, Alexander C. Humphries rebutted Taylor's previous remarks about the failings of engineering education in the matter of providing top-quality, innovative engineers, by noting that "not all of the men will be able to go to the top."

and for the young men each was bringing through either shop or school.

This categorization of engineering elite versus engineering educators does not explain the actions of all individuals. There were many who did not fit into this classification system and who did not hold the appropriate opinions. Nevertheless, this division explains many alignments among engineers that are not easily explained by any other conceptual framework. Further examples can be found in the debate over the relative merits of school and shop.

Apprenticeship was a key issue. Elite and educators alike usually lamented that the old apprentice system was disappearing because of the routinization of shop work and the lack of interest on the part of manufacturers. It was true that mass-production techniques had reduced the importance of apprenticeship as a training tool. It was easier to let a man stay with a task he had mastered than to provide broad experience at great cost to the establishment. The elite mourned the decline of the system because it was to them the real period of acclimitization to the shop; the educators mourned its death because such public mourning gave a rationale for the expansion of technical education at *all* levels to fill the ranks of industry at *all* levels.

The truth of the matter was that apprenticeship was not dead at all in the traditional industries in which mechanical engineers worked. The editors of the *American Machinist* in 1896 surveyed a number of leading firms to establish the status of apprenticeship (see Table 2). Of the 37 builders of stationary and marine engines, pumps, etc., surveyed, 27 took apprentices, 3 had discontinued the practice, and 24 found the system quite satisfactory. The figures for the 35 machine-tool builders were respectively 25, 1, 24; for the 25 locomotive builders and railroad companies, 22, 1, 20; and for the 19 builders of miscellaneous machinery, 11, 2, 10. These figures, while indicating nothing about the *quality* of the apprenticeship programs, do indicate that the system was still very much in use and that there was no trend toward its elimination.[21] Even as late as

21 "Status of Apprenticeship in the Trades Concerned in the Production of Machinery," *Amer. Mach.*, December 24, 1896, pp. 1183–202.

1908, when the journal printed a letter from a boy asking how to become a mechanical engineer, the replies from readers ran as follows: 75 per cent favored apprenticeship before the college course and only 25 per cent recommended the reverse. All agreed that apprenticeship in a shop was necessary at some time in the process of becoming a mechanical engineer. The institution was far from dead.[22]

TABLE 2: Status of Apprenticeship in 1896

Type of Concern	Number of Replies	Number Taking Apprentices in 1896	Apprenticeship Discontinued	Found System Satisfactory
Builders of stationary and marine engines, pumps, etc.	37	27	3	24
Builders of machine tools	35	25	1	24
Locomotive builders and railroad companies	25	22	1	20
Builders of miscellaneous machinery	19	11	2	10

Why then did so many fear that the apprenticeship system was dying? The elite thought so because structural changes born of mass-production manufacturing were beginning to find their way into the smaller shop world and were beginning to destroy the close personal relationships that mechanical engineers had traditionally had with the men of the shop. As long as both made their way through the same apprenticeship system, this personal contact led to understanding and respect between the men of office and shop. The mechanical engineer always moved on to the office, but he could take with him the feeling that he had (whether he did or not) a deep knowledge of the workman and his work. John C. Hoadley, an innovative engi-

22 "How Should a Boy Go about Becoming a Mechanical Engineer," *Amer. Mach.*, December 17, 1908, pp. 900–1. There was feeling that the quality of apprenticeship had declined. See a talk by John Edson Sweet printed in Albert W. Smith, *John Edson Sweet* (New York: American Society of Mechanical Engineers, 1925), pp. 99–116.

neer-entrepreneur in the steam engine and machine shop business reminisced in 1881 about his early days as a mechanical
engineer, particularly his experiences in 1847 when working in
a shop as a civil engineer, but with the duties of a mechanical
engineer.

One evening after bell time, three or four apprentices came
to his office and timidly asked if Hoadley would give them
lessons in mechanical drawing. An agitator for the ten-hour law
had told the apprentices to expect rebuff since "Mr. Engineer
knows the value of an education, and won't be such a fool as
to cheapen his own attainments by sharing them with you."
Hoadley responded by setting up an after-hours class for about
a dozen young men, many of whom later held responsible positions and each of whom he counted as a personal friend.[23] To
men like Hoadley the situation of a mathematically oriented
young technical graduate with no experience barging into the
shop and clashing with its culture rather than blending with it
would seem a travesty on a fine institution.

If the engineering elite was disturbed by the fact that the
technical graduate was frequently unwilling to go through the
shop experience as an apprentice and to learn humbly from the
tobacco-chewing, rough old hand, the new manufacturing industries which required the services of these graduates were in
no mood to quibble over lost relationships. By the mid-nineties
a number of firms, particularly the large electrical equipment
companies such as General Electric and Westinghouse, had
developed regularized and systematic programs for giving technical graduates practical shop experience without exposing them
to the rough and tumble shop culture world. It was the final
step in the process of superseding shop culture as an educational
and professionalizing institution.

The technical graduates hired under these programs were
paid low wages for the first two years and were assigned for
six months to the drafting room, six months to the testing departments, six to the manufacturing division, and so on. At all

---

23 J. C. Hoadley, "Engineering Reminiscences," *Amer. Mach.*, January 8, 1881, p. 5.
Hoadley's family came to America in 1663. His father was a fairly well-to-do farmer
in Lewis County, N.Y. Frank A. Taylor in *DAB* s.v. "Hoadley, John Chipman."

times it was recognized that these were special employees, who were being trained for leadership positions in the engineering department and for management positions generally. The purpose was to give the technical graduates an understanding of the processes in each department, not to force them to compete with the wage-earning machinists on an equal basis or even to encourage them to become expert workmen themselves.[24]

The electrical industries were the pioneers in this sort of program, but some of the large railroad shops were also running special apprentice programs in an effort to attract technical graduates, and by 1910 other types of firms, such as Allis-Chalmers, offered similar arrangements.[25] The professors in the engineering schools approved heartily, since the system provided for orderly and regularized entry into professional activity by their graduates. As an engineering instructor wrote to the *American Machinist* in 1907, "all must do the same thing that the big electrical companies do—take our graduates as they are, and spend a year or two giving them an insight into their particular branch of the business."[26]

The mechanical engineering elite, as might be expected, disapproved of these programs since their net effect was to circumvent and destroy the true value of shop culture. Roy V. Wright—who had a technical degree from Stevens, worked his way through the railway shops, became a leading technical journalist and editor for the railway engineering industry, and was president of the ASME in 1931—expressed relief in 1909 that some railroads were abandoning the special apprentice system. Such programs did not enable the potential mechanical engineer "to get into close contact with men in the ranks and this practice also discouraged the regular apprentices, for it practically served notice on them that the best opportunities were open to the specials. It is doubtful whether the system as it existed, benefited either the regular or the special apprentices,

---

[24] Thomas Howe, "The Training Course for Engineers of the General Electric Company," *Sibley Journal*, reprinted in *Wisc. Engr.*, VI (February, 1902), 173.

[25] G. J. Kircher, "Engineering Student Apprenticeship," *Minn. Engr.*, XIX (January, 1911), 86.

[26] Instructor, "The Technical Graduate from the Professor's Standpoint," *Amer. Mach.*, January 31, 1907, pp. 152-53.

and on most roads it has been done away with, the men being taken in and advanced strictly on their merits."[27] The special apprentice did not have to face the uncertainty of shop culture.

Another area of conflict between the advocates of school and shop was the question of admission standards to the engineering schools. After the doldrums of the seventies and eighties had passed, mechanical engineering education experienced a great increase in students, and the schools were enabled to increase admission standards accordingly. In some cases they were forced to do so to avoid seriously overtaxing their facilities and faculty. Most of the emphasis in the examinations and admission requirements was on knowledge of subjects which were taught in the high schools, both the older, private preparatory high schools and the newer, state-supported public high schools. By 1892 a survey showed that 49 per cent of the country's engineering schools required physical geography, 48 per cent English literature, 40 per cent English history, 80 per cent American history, 82 per cent algebra, and 86 per cent plane geometry.[28] Engineering educators saw this as a welcome development. It made possible the preselection of those with aptitude for learning in the school context and it provided an assumed base of knowledge upon which the technical professional school could enlarge. To criticisms from the shop culture enthusiasts that such admission requirements denied access to the boys from the shops who had never been to high school, educators like Robert H. Thurston replied that men from the shops were welcomed provided they could master higher mathematics.[29]

American Machinist, frequently representing on its editorial pages and elsewhere the defender of shop culture, attacked high standards in 1896 on the following grounds: "Instead of testing the abilities of the student, their real effect is to test his opportunities, and to sift out those who by reason of location, occupation, or lack of means, have been denied access to the best

---

27 Wright, "The Mechanical Engineer and the Railroads," p. 140.
28 Robert H. Thurston, "Technical Education in the United States," ASME, *Trans.*, XIV (1892–93), 982.
29 Letter to the Editor, "Entrance Requirements, Endowment and Scholarship at Cornell University," *Amer. Mach.*, March 5, 1896, p. 267.

preparatory schools. . . . The entrance requirements could not be better if they were intended to exclude students from the shops."[30]

Both of these lines of argument depended upon whether shop experience was gained before or after college training. The engineering educators, interested in having the first and most permanent influence upon the ideas and attitudes of the potential mechanical engineer, rejected the idea that shop experience should come before technical training. The route they favored was common school, preparatory or high school, possibly a general college education, then technical or professional education. Any shop experience could be obtained afterward, either in the traditional apprenticeship way, or preferably, in the special programs set up in some industries. The mechanical engineering elite, however, frequently disagreed, believing that shop experience was necessary to make the mind ready for learning of a formal variety. Alexander Lyman Holley outlined this position in 1876 before the American Institute of Mining Engineers. He argued that engineering schools were laying too much emphasis on abstract principles, and he advised putting actual practice first. The debate which followed his talk led to a joint meeting of the American Society of Civil Engineers (ASCE) and the American Institute of Mining Engineers (AIME) in Philadelphia during the Centennial Exhibition.[31] Agreement was reached that practical training was necessary *early* in the professional life of the engineer. By the 1890's the engineering educators disagreed sharply.[32]

Related was the question of whether the mechanical engineer

[30] "High Admission Requirements for Engineering Schools," *Amer. Mach.*, February 13, 1896, p. 196.

[31] Mavis, "History of Engineering Education," pp. 192–93. Also sparked by the Centennial was interest in the Russian system of engineering education which was displayed there. This system required those training for engineering to become manually skilled before they could proceed with more abstract scientific training. See *Description of the Collection of Scientific Appliances Instituted for the Study of Mechanical Art in the Workshops of the Imperial Technical School of Moscow* (Moscow: W. Gautier, 1876), pp. viii–xix.

[32] For contrasting opinion on this, see "Discussion on Paper No. 850," pp. 646–767; Woodward, "The Training of a Dynamic Engineer," pp. 742–83; Jackson, "College and Apprentice Training," pp. 473–530; Alden, "Technical Training at the Worcester Free Institute," pp. 510–65; and "Technical Education," *Amer. Engr.* (Chicago), II (January, 1881), 6–7.

had to have technical proficiency as a machinist, mechanic, and workman or merely needed to have passing familiarity with the processes involved in shop work. Almost all of the educators believed that it was not necessary for the mechanical engineer to be proficient in manual shop skills, but the elite frequently expressed the opinion that he should be. "Instructor" wrote the *American Machinist* in 1906 that "it seems to me that while manual dexterity is unnecessary to a student who wishes to become an engineer, he cannot be too well grounded in the multiplicity of shop processes."[33] Professor G. H. Shepard of Cornell's Sibley College stated this concept even more strongly in 1900: "I submit that the engineer who has served his time as a machinist has no advantage over the engineer who, with less knowledge of the machinists' trade, combines more knowledge of the other engineering trades."[34]

The Worcester Free Institute, which attempted to recreate the shop culture and demanded excellence as a machinist for graduation, appealed to many of the mechanical engineering elite. Over the years a number of papers were presented before the ASME on the program at Worcester by its instructors George I. Alden, M. P. Higgins, and Ira Remsen. Their case, seconded by many engineers, was, as Remsen expressed it, that contact with the things of nature was more important than "contemplation in a study. . . . Manual labor is necessary. Without it we may as well give up hope of acquiring knowledge of the truth."[35] John Edson Sweet believed that the Worcester people were heading in the right direction and that "every engineer ought to be a workman."[36] Even the scholarly, sometime professor Coleman Sellers found it possible intellectually to appreciate the idea of a mechanical engineer who was not an expert machinist and craftsman, but he could not emotionally believe it.[37]

33 "The College Shop," *Amer. Mach.*, June 21, 1906, pp. 813–14.

34 "Discussion on Paper No. 850," pp. 701–2.

35 Alden, "Technical Training at the Worcester Free Institute," pp. 510–65; M. P. Higgins, "The Education of Machinists, Foremen, and Mechanical Engineers," ASME, *Trans.*, XXI (1899–1900), 1114–53; Ira Remsen, "Thinking Unprofitable without Continual Investigation and Experiment," *Amer. Mach.*, August 18, 1904, p. 1090.

36 "Discussion on Paper No. 850," p. 721.

37 Coleman Sellers, "President's Address, 1886," ASME, *Trans.*, VIII (1886–87), 680–81.

Robert Thurston, however, believed that Worcester was a first-class trade school but "not a professional engineering school."[38] Dexter S. Kimball, teaching at Cornell at the turn of the century, believed "It should not be necessary for an engineer to be *able to actually do* all the processes in a shop"; he need only have enough knowledge to *design* in a knowing manner.[39] Some educators and others were willing to admit that it would be a good idea for a mechanical engineer to be an expert machinist, but even these were unable to offer any way that such training could be added to the technical curriculum without keeping the student in school an inordinately long time.

Given the conclusions that the mechanical engineering educators reached regarding the role of shop training and its importance, one might suppose that they regarded the mechanical engineer as something apart—different in kind, not merely in degree—from the mechanic and machinist. Likewise, one might expect the mechanical engineering elite to see the profession as a continuum with the more or less talented individuals occupying the different rungs on the ladder of position and status. Such different opinions did develop.

M. P. Higgins, speaking for Worcester Free Institute, told the ASME in 1900 that the mechanic was the universal man of the hour, of which the engineer was only an advanced type. The difference between the machinist and engineer was one of "degree and not of kind."[40] This was, he admitted, opposed to the current trend in technical education. *American Machinist* commented on Higgins' paper that "there are some who will probably resent his grouping of the education of machinists, foremen, and engineers together," but the editors felt the difference was in degree only.[41] Shop defenders like Fred Halsey, H. Wade Hibbard, and F. R. Hutton agreed.

Robert Thurston, also commenting on Higgins' paper, summed up the case of the engineering educators when he spoke of the good work Worcester was doing training the "machinist

---

38 "Discussion on Paper No. 850," p. 698.
39 *Ibid.*, pp. 702–3.
40 "The Education of Machinists, Foremen, and Mechanical Engineers," p. 1122.
41 January 4, 1900, p. 1.

. . . for the purposes of the engineer."[42] They made a mistake, he told the ASME, when they did not fully recognize that the training for a mechanical engineer must be distinct from that for a machinist. David S. Jacobus, another engineering teacher, pointed out that some of his graduates who were very poor in the shops made "striking successes in their chosen fields of work," indicating that the training and the work were of a different kind than that for machinists.[43] Professor C. H. Benjamin differentiated between "thinkers" and "workers" and claimed that the demand for thinkers, the mechanical engineers, was high.[44]

Discussion of the creative and design role of the mechanical engineer brought up quite naturally the question of whether intuition (a convenient word for spontaneous creative intelligence) was an inborn quality or one which could be learned. Shop engineers tended to agree with *Engineer*'s editorial of 1894 that "engineers are born, not made, but the born engineer seldom comes out of the technical colleges. Oftener he comes out of the woods."[45] The educators were more likely to feel that intuition and mechanical ability could be taught, though they rarely specified how this was to be done. The concern was really one of whether mechanical engineering was to be an elite or a mass profession.

The educators developed an interest in the student, particularly the average-to-good student, and oriented their courses and programs around him. An editorial in the *American Engineer* stressed that it was indeed the young man of average ability who had the most to gain from technical education.[46] Dodge's 1903 study of the money value of technical training stressed that, while a few untrained geniuses might make it to the top, technical training could raise the income potential of the "average man."[47] Many of the educators' statements were cast in democratic ideology, that the free or low-tuition schools

42 "Discussion on Paper No. 850," p. 695.

43 *Ibid.*, p. 741.

44 Comment on Higgins, "The Education of Machinists, Foremen, and Mechanical Engineers," ASME, *Trans.*, XXI (1899–1900), 1135.

45 "Graduates of Engineering Colleges," *Engineer* (New York), November 10, 1894, p. 114.

46 "The Education of the Engineer," *Amer. Engr.* (Chicago), November 24, 1882, p. 243.

47 "The Money Value of Technical Training," pp. 40–48.

would make it possible for anyone to find opportunity through education and to increase occupational mobility.

William F. Durfee represented the reaction of many of the shop elite in his skepticism that "reasonable intelligence, money, and three or four years of engineering school" could make a true mechanical engineer.[48] Engineers were born, not made. Some were led by such thoughts to a feeling that perhaps the trouble had come when technical education first broke away from the confines of classical education. One can imagine that the older mechanical engineering elite from the best eastern families would have had little indeed to talk about (besides friction of bearings, and so on) with an inexperienced technical graduate from a midwestern school. Perhaps, said John R. Freeman to the assembled students and faculty of Case Institute in Cleveland in 1905, "the old academic education fits better for the position where one deals with men, or for the $10,000 position, while the technical school fits better for the position that deals with materials, or for the $4,000 position."[49] He also stressed the importance of being socially able as well as highly proficient at scholarship. Freeman had a technical education, was an innovator in hydraulic machinery and fire prevention, a consultant on the building of the Panama Canal, and ASME president in 1905.[50]

Sometimes the elite expressed this concern, as did ASME president Horace See in 1888, in terms of the need for schools to develop the whole man, not just cram the memory.[51] Oberlin Smith[52] is interesting because he was a member of the shop elite who walked the tightrope between extreme faith in the promise of technical education and the old concept of shop culture. Smith, a technical graduate himself, and chief engineer of a machine tool company in southern New Jersey, called for "The Engineer as a Scholar and a Gentleman" in his presidential address in 1890. Smith told the ASME that "all our engineers cannot come from cultivated homes," and that class distinctions

48 Comment on Thurston, "Technical Education in the United States," ASME, *Trans.*, XIV (1892–93), 1008.

49 "A Plea for Breadth of Culture in the Technical School," *Amer. Mach.*, January 12, 1905, p. 64.

50 ASME, *Jubilee Book*, p. 34; Hutton, *History*, pp. 114–15.

51 "President's Address, 1888," ASME, *Trans.*, X (1888–89), 488.

52 ASME, *Jubilee Book*, p. 43.

81

Machining operations on rotary pressure blowers at the P. H. & F. M. Roots Co., Connersville, Ind., 1880 (*Scientific American*).

were disappearing. He cautioned the members not to oppose the advance of technical education because such a position could lead them to scoff at theory and science. Nevertheless, he called upon all mechanical engineers to develop the ability to "meet [their] wealthy and cultivated clients upon their own level; to excel them, if anything, in the intelligent appreciation of their mutual affairs, and in the amenities of social intercourse." Smith's vision was appealing to school and shop supporters alike, but it was unrealistic.[53] He was asking the technical schools to produce not merely engineers, but a select corps of engineers able to mix socially with the other professional elites and with the upper social class. He was asking them to do this when the numbers of engineering students were rapidly moving into the thousands.

American engineering educators did not react to such challenges in a uniform way, and this is one area where the school-versus-shop analogy does not seem to fit. It fits with Robert Thurston and a few others who maintained that the technical school must not get involved in general education. Some engineering educators believed by the early twentieth century that a broad, general education was essential in connection with technical education. Alexander C. Humphries (until 1902 a consulting engineer in gas plant work) dedicated himself, as Stevens Institute's new president in 1902, to broadening the Institute's educational base to include English literature, logic, history, economics, and modern languages.[54] As Wisconsin's Dean of Engineering F. E. Turneaure wrote in 1905: "The value of a general college education as a preparation for law, medicine and theology has long been recognized. As a foundation for the engineer's training it is quite as valuable." Turneaure believed that those who had a general education would have a far better chance of success than those with merely technical training.[55]

General education was not an issue on which shop engineers agreed either. Frequently opinion divided along lines of prior education rather than along lines of allegiance to shop culture,

[53] Oberlin Smith, "President's Annual Address: The Engineer as a Scholar and a Gentleman," ASME, *Trans.*, XII (1890–91), 42–55.

[54] "Broader Training for the Engineer," *Amer. Mach.*, January 29, 1903, pp. 153–54.

[55] "The Engineering School in the United States," p. 275.

and few could dispute the schools' claim to pre-eminence in the domain of general or classical education—here was one area where shop culture could not supply the needs of the mechanical engineer.

Because of practical difficulties (such as lack of time and funds) few engineering schools before 1910 developed extensive general education for majors in mechanical engineering. Those "extra" courses which were added, such as accounting, statistics, practical English, and economics, were justified primarily on the practical grounds that they helped an engineer to earn a living in an industrial world in which he was frequently at a disadvantage in competition with men of business ability.

There was one significant attempt to bridge the gap between school and shop culture. Herman Schneider was the best-known proponent of cooperative education, as it was called in his program at the University of Cincinnati and elsewhere. Schneider's plan was to make arrangements with local manufacturing industries so that young men in the mechanical engineering course at Cincinnati could spend part of each day, week, or month (the system was tried in all forms) in the shop. This would hopefully end forever the controversy over the proper order of shop and school by making them coterminus. Though it attracted some interest among shop enthusiasts, the top educators could not be enticed by the system. It was unlikely to become a significant movement in engineering education for a number of reasons. Manufacturers were disinclined to have different men on machines at different hours or days, or to have machines actually idle part of the day; then too, such an arrangement would be nearly impossible in nonindustrial areas such as Madison, Wisconsin, or Ithaca, New York, both locations of eminent engineering schools; and the single most important factor against Schneider's plan was simply that the educators were not interested in it.[56]

---

[56] "Notes on the Cooperative System," *Amer. Mach.*, July 28, 1910, pp. 148–51. Shop culture advocates were fascinated with the idea. Frederick W. Taylor made a special trip to Cincinnati in 1909 to study the system and wrote that "the general principle of this approach appeals to me as most admirable." Taylor to Calvin W. Rice, February 12, 1909, in Taylor Collection.

The correspondence school movement was another possible answer to the problem of shop versus school. It originated from the desire of people to get a technical education while working at a full-time job, and it grew from courses that prepared stationary engineers to pass exams to full-fledged courses in mechanical engineering and many other technical subjects. It never was able, however, to gain support from the elite of engineering education or practice.[57]

[57] For good examples of correspondence school advertisements, see *Scientific Machinist*, especially October 1, 1896. The reaction of engineering educators to the phenomenon can be seen in Victor C. Alderson, "A Plea for Technical Education for the Mechanic," *Amer. Mach.*, August 14, 1902, pp. 1173–74.

The clearest way to show the complexities of the school-versus-shop controversy in mechanical engineering education is to examine Sibley College of Engineering at Cornell University in Ithaca, New York, which had, in succession, leading proponents of each culture, John Edson Sweet and Robert Henry Thurston.

The school of mechanical engineering at Cornell University began like many such schools at land-grant colleges. The Mechanic Arts Department, headed by John L. Morris, existed from the mid-sixties when the university first began taking students, but it lacked either clear purpose or wide support among the trustees. Cornell itself was, like Purdue, a mixed product of land-grant funds and private funds of a major donor, in this case Ezra Cornell, self-made telegraph magnate. One trustee, Hiram Sibley, who shared the same self-made background as Cornell, became interested in mechanic arts training and in 1870 pledged a sum of money to erect a building and create a more extensive program.[1]

There was some early confusion over whether the institution was training mechanics or engineers or both. When Hiram Sibley's gift bore fruit in 1871 as Sibley College of the Mechanic Arts, the degree of Bachelor of Mechanical Engineering was promised to students who graduated from the course. By 1873 the services of John Edson Sweet had been procured for the position of master mechanic and director of the machine shops.[2]

Sweet saw himself as a missionary to the pagan, and he wrote for the technical press in 1873: "I have been placed in charge of the practical department, or in other words, delegated to harmonize the book and brain science, on the one side, with the actual hand practice, on the other."[3] He was a most en-

---

[1] Morris Bishop, *A History of Cornell* (Ithaca, N.Y.: Cornell University Press, 1962), p. 169.

[2] ASME, *Jubilee Book*, pp. 43–44; Hutton, *History*, pp. 13, 85–88.

[3] "Plain Talk," *Industrial Monthly*, II (April, 1873), 72.

thusiastic supporter of shop culture, as he told a reporter in 1899: "If you want to see ingenious men and rigs, go into the country shops."[4] His idea of education was a master craftsman surrounded by humble, eager, admiring apprentices, learning about mechanics, industry, and life. He hated the routine and standardized nature of much educational practice, and spoke of "examinations and diplomas" as the two great evils of schools and colleges. The ideal university or college would provide a good shop where the students could turn out high-quality commercial work for eight hours, and then study four hours each night.[5] It should be, even more than the program of the Worcester Free Institute, an attempt to recreate shop culture in a school situation.

Sweet's attitude, that it is not possible to deal with everything on a theoretical plane, was expounded in his best-known paper, given before the ASME in 1885, "The Unexpected Which Often Happens." It was not, as it might appear to the casual observer, an attack on science or theory per se, but rather a series of observations of cases in which theory did not immediately or perfectly explain the strange results. It was a plea for recognition of the importance of uncertainty in any endeavor and a plea to educate the young to appreciate that kind of uncertainty and to be able to deal with it. As he concluded:

It is not always the uneducated, the insane or the stupid who produce failures, nor the best educated, most thoughtful, or most experienced who bring out everything according to the original intention. The unexpected comes to good and bad alike, and so in our teachings to the young and our planning for ourselves, is it not well to have our statements and our speculations pretty well saturated with the elements of uncertainty?[6]

This talk struck a responsive note among elite defenders of shop culture. Henry R. Towne, a college-trained engineer (University of Pennsylvania and the Sorbonne), who had spent some time in the William Sellers shops, became a pioneer in mass-production technique, profit sharing, and scientific management, and four years later was a president of the ASME,

---

4 "A Talk with Professor Sweet," *Engineer* (N.Y.), January 15, 1899, p. 16.
5 John E. Sweet, "An Industrial University," *Amer. Mach.*, March 12, 1891, pp. 1–2.
6 ASME, *Trans.*, VII (1885–86), 156.

contributed several examples from his own experience. So did Oberlin Smith and William F. Durfee. The argument was against shoddy science, science that claimed too much and was too sure of itself.[7]

Little is known about the actual content of Sweet's course other than that it was eminently practical training conducted in the shop, not the classroom. He acquired a small but devoted group of students, who literally worshipped him. He apparently made little or no attempt to communicate with the Board of Trustees or the university administration. Neither did he make any effective attempt to increase the faculty, facilities, or resources of his department. When appropriations were not forthcoming for equipment he needed, Sweet became disgusted and quit in a huff in 1879.

By 1882 *American Machinist* noted that of the forty-four graduates from Sibley College since 1869, forty-one had completed their work up to and including 1878, but since that time only three had gotten degrees and none had that year. The editors reported that they had heard that Sweet *was* the department and that his leaving had impoverished the program.[8] Cornell's president Andrew Dickson White made an effort to answer such charges in the technical journals by pointing out that Sibley College had two professors and much new equipment.[9] Nevertheless, as soon as Sweet quit and for six or seven years afterward, trustees, and later alumni, joined with the administration in seeking a top-flight engineer to replace him and to raise the foundering program.

Complicating the matter of a replacement for Sweet, and perhaps even one of the reasons for his leaving, was the controversy surrounding Professor John L. Morris. Sweet had apparently clashed with Morris in the mid-to-late seventies over the control and administration of the machine shops. Sweet's solution, which he presented to the executive committee of the

---

7 *Ibid.*, pp. 152–63. Henry R. Towne belonged to the ninth generation of a Salem, Massachusetts, family that came to America in 1640. He was educated in private schools. His father provided the financing for the lock company he started with Linus Yale. Carl W. Mitman in *DAB* s.v. "Towne, Henry Robinson."

8 *Amer. Mach.*, July 15, 1882, p. 8.

9 *Ibid.*

trustees in 1878, was a simple one: that he be made full professor of practical mechanics and director of the machine shop and that the Department of Mechanic Arts be divided so as to give complete control of the shop and all practical work to Sweet. Morris apparently acquiesced unwillingly, and the change was made. Even the new arrangement was not satisfactory to Sweet, who resigned in 1879, but not before a student attempt to oust Morris had flourished and been stifled.[10]

On several occasions in 1878 a group of students presented to the Board of Trustees of the university a petition outlining in detail why Morris should be fired for incompetence. The petition, the letters, and other documents the students submitted over the next year all praised John Edson Sweet very highly but also advocated an expansion of the program and a tightening of standards.[11] Morris, they contended, was "worse than nothing."[12] It is not surprising to see Fred Halsey, Cornell '78, attacking an incompetent professor or praising the first man of shop culture. Albert W. Smith, another petitioner, went on to become both a mechanical engineering professor and an interim president of Cornell. Arthur Falkenau was another, and he later became president of the exclusive Engineers Club of Philadelphia. One of the founding members of the ASME in 1880 was John Saylor Coon, another petitioner, a graduate of 1877, who had stayed on for an extra year as a sort of graduate assistant in the department. The critics of Morris were, even as they criticized, nascent members of the mechanical engineering elite.

Halsey, Falkenau, and Coon were the ringleaders and mainstays of the protest movement, but they enlisted others in their cause. Edward N. Trump, class of '79, contributed stories of Morris' incompetence. Trump later became a vice-president of the ASME and an innovator in processing and management techniques. A. W. Smith indicated his agreement when he wrote of Morris to the university's acting president Russel: "His in-

---

10 Cornell University, Executive Committee Minutes, October 14 and 30, 1878, Cornell University Archives, Ithaca, N.Y. (hereafter cited as Ex. Com. Minutes).

11 Cornell University, Board of Trustees, Papers, Cornell University Archives, Ithaca, N.Y. (hereafter cited as Trustees' Papers).

12 Letter to Professor W. C. Russel, April 20, 1878 (signature of sender erased), in Sibley College Papers.

competence disgusted me." If he stayed at Cornell, Sibley would become "the laughing stock of all men in technical education."[13] The petition itself accused Morris of holding ideas which were common among "ignorant mechanics."[14]

John Saylor Coon wrote to Andrew Dickson White in 1879 that he was able to get a job with the firm of E. D. Leavitt, Jr., as first assistant only through the special offices of John Sweet. His mechanical theory training from Morris had been so poor that, had he not done much private studying, he would have lost the position through incompetence. Several of these men had, like Coon, not only studied on their own, but had come to technical education from the shop rather than directly from high school. They had come probably because of the magic of Sweet's name and in the correct belief that Sweet could facilitate their re-entry into shop culture after their technical training.[15]

The protest group detested Morris because they saw him as the incompetent, impractical academician, hanging on for dear life to a sinecure, hoping to disguise his incompetence with bravado and evasion. Whether Morris was actually incompetent is not relevant here. It is clear from the comments of the students and alumni that even if Morris had been competent in dealing with pure mechanical theory, it was his complete ignorance of and lack of interest in mechanical engineering practice that repelled them. The incompetence which they felt characterized Morris' theoretical work merely compounded the original sin of ignoring shop culture. The students were objecting not to science, however, but to science presented without a hint of possible application.

Edward B. Gardiner, engineer of the Newark City Silk Mills and a Cornell graduate, wrote to President White in September and October of 1878 to offer concrete suggestions as to how

---

13 Undated letter, in Sibley College Papers.

14 In Trustees' Papers.

15 January 30, 1879, in Sibley College Papers. This group and some others who worked for Sweet at one time or another called themselves Professor Sweet's "boys" and for years met for annual dinners with Sweet as the honored guest. A. W. Smith, *Sweet*, pp. 74–79. See pp. 60–61 for an account of the reasons Edward N. Trump had for going to Cornell to work under Sweet, how he became one of Sweet's "boys," and his later rapid rise aided by Sweet's recommendations.

the program at Cornell might be improved. Cornell was, Gardiner wrote, producing mere draftsmen and not stressing such important subjects as strength of materials, economy of steam engineering, laws of condensation and expansion, and transmission of power. In the industrial world of 1878, "the mechanical engineer is taking a more special and higher position, than ten or fifteen years ago, and the draughtsman has become a recognized unit, a little higher than the machinist and a good deal lower than the engineer." Gardiner saw the problem objectively, in the context of recent and inexorable changes in the role of mechanical engineers in industry.[16] He realized the need for specialization of training and for realistically appraising the needs of industrial society. He wrote to E. D. Leavitt to ask who would make a good replacement for Sweet and got the reply that Alexander Lyman Holley, Charles T. Porter, J. Coleman Sellers, or S. B. Whiting had ability; all of them had great respect for shop culture and were part of it.[17]

Some of the Morris-Sweet controversy can be attributed to personal characteristics and to a struggle for control within the department. Morris thought Sweet was a pettifogger and insane to boot.[18] Sweet was convinced Morris was incompetent and worse than useless as a teacher. Each gathered about him a corps of loyal students and later of alumni. This would all be just a bundle of dirty laundry not worthy of being aired if it were not that the details of the controversy tell us something about the development of technical education. Morris, of course, denied all the charges and made countercharges, such as that Halsey was a liar and Coon was lazy.[19] He marshaled his loyal crew of graduates and promised the trustees a sheaf of adulatory letters.

What he got was sympathy from former students, who affirmed Morris' ability as a teacher, but particularly stressed his qualities as an individual. In 1878, and throughout the next

---

16 September 16 and October 12, 1879, in Sibley College Papers.

17 Leavitt to Gardiner, undated, in Sibley College Papers. The importance of a recommendation from Leavitt is made clear in a letter from Washington Jones to Coleman Sellers, April 16, 1890, in Peale-Sellers Papers.

18 See especially Morris to Andrew D. White, July 26, 1879, in Sibley College Papers.

19 Morris to White, October 12, 1878, in Sibley College Papers.

six years of the controversy, the pro-Morris letters came from men who were from the West and who held jobs as draftsmen or less. By contrast, as we have seen, the anti-Morris faction was eastern-oriented and became, almost to a man, part of the mechanical engineering elite. The anti-Morris group was full of suggestions on how to improve and change technical education at Cornell whereas the pro-Morris faction was dedicated to the proposition that all was well and no change was needed. Morris was neither a competent educator nor a functioning member of the shop elite. To place graduates in 1878, it was necessary to be one or the other. Morris' most enthusiastic supporter was a draftsman for the Chicago, Burlington, and Quincy Railroad.[20]

The university administration was caught in a difficult position. They were afraid to fire Morris because of the policy implications in making faculty competence subject to student judgment. The Executive Committee of the Board of Trustees tried to placate the Halsey-Coon-Falkenau triumvirate by a proposal to send Professor Morris to Europe for a year of study and travel. Falkenau, Smith, and W. K. Seaman replied that such a whitewashing of the affair would do no good.[21] "Can you pretend to say . . . that a man of Professor Morris' age and indolent habits can become a mechanical expert in one or two or three years?" asked Seaman. Seaman was fully within the elite shop-culture tradition: he was working for A. L. Holley in New York in 1879 when he wrote to Acting President Russel that he might expose the whole affair to the "public press."[22] Halsey thought the send-Morris-abroad plan was like "treating a case of small pox or yellow fever with catnip tea." Diplomacy, not statesmanship, he felt, had characterized the whole affair.[23]

The trustees and the administration, though obviously under great pressure, stuck to their decision to re-educate Morris, but

---

20 For examples of opinion favoring Morris, see letter from W. Thackrey (signature unclear), October 14, 1878, in Sibley College Papers; A. B. McNairy, C. W. Mason, and H. A. McDenwid to Trustees, June, 1884, in Trustees' Papers; J. S. Waterman, B. Johnson, and R. H. Treman to Andrew D. White, June 18, 1884, in Trustees' Papers; Charles E. Lipe to President and Board of Trustees, June 16, 1884, in Trustees' Papers.
21 To White, dated 1878, in Sibley College Papers.
22 May 9, 1879, in Sibley College Papers.
23 Halsey to W. C. Russel, May 14, 1879, in Sibley College Papers.

shifted the emphasis from his fate to the more basic question of what Sibley College was to become in the years to follow. After the resignation of Sweet and the departure of Morris for Europe they tried a one-year appointment of a Philadelphia man, Samuel G. Powel.[24] Andrew D. White, cognizant of the blow the university would suffer if Sibley College lost prestige in the engineering world, searched through 1879 to 1880 for a true successor for Sweet, hopefully one who would satisfy both school and shop devotees. By July, 1880, he thought he had found one in J. Burkitt Webb, then studying in Germany.

Webb, born in Philadelphia, entered the classical course at Philadelphia High School, took a fling at experiments in electro-magnetism and medical publishing, and picked up some drafting training in night school at Philadelphia Polytechnic University. With a friend he probably met at the university, Oberlin Smith, he founded a machine, die, and tool company in 1863 in Bridgeton, New Jersey. He was thus thoroughly familiar with shop culture and had wide experience before he entered the University of Michigan in 1868 and sought an engineering degree under De Volson Wood. After a postgraduate year as Wood's assistant, Webb was called to the Chair of Civil Engineering at Illinois Industrial University where he worked closely with S. W. Robinson. He felt called in 1878 to go to Europe for further study and apparently had the means to do so since he was able to resign his position in 1879 to spend another year of study in Göttingen and Heidelberg, followed by a year in Paris.[25]

In 1880 Cornell President Andrew D. White asked him to prepare a set of recommendations for improving engineering training at the university. His suggestions included adding higher mathematics, through calculus, and French and German for the first two years.[26] White was impressed and secured Webb's appointment to the Chair of Applied Mathematics and Theoretical Mechanics in 1881. Webb was obviously in Morris' and not Sweet's area and Morris resented the intrusion. Spurred

24 Ex. Com. Minutes, September 6, 1879.
25 Webb to Andrew D. White, July 1, 1880, in Sibley College Papers.
26 Webb to White, July 7, 1880, in Sibley College Papers.

on by his 1878 victory, Morris pressed White in 1883 for control of curriculum and appropriations. Webb had reorganized the curriculum and pushed Morris aside, and Morris asked to be retained as "Dean of the Department, to be recognized and sustained as such." Webb was to be demoted and placed on the same footing as the other professors. Morris apparently had his way.[27] Though Webb stayed on until 1886, he never regained control.

By May of 1883, student and alumni protest was rising again. The students published their criticisms in a small daily paper which was quickly suppressed. The alumni held meetings in New York City, and the *American Machinist* reported a consensus that six to eight years earlier the mechanical engineering school had been one of the best, but now it was not even good.[28] Halsey, Coon, and Falkenau were not to be left out of the fray. Now highly successful alumni, the three—from a 23 Park Place address in New York City—sent a long, printed petition in which they called for immediate action and compared Cornell's engineering education invidiously to that at Stevens Institute of Technology. They particularly noted that the machine shop was two and a half times larger at Stevens than at Sibley. The trio demanded Morris' retirement, appointment of at least one professor with both scientific and practical experience, and increasing expenditure for equipment under the direction of such a professor. Also among their recommendations was the appointment of nonresident professors "from among the prominent engineers of the country."[29]

The trustees and administration of the university had, by June of 1884, decided that someone must be found to reorganize the whole program. A committee was appointed, consisting of White, Chairman of the Board Henry W. Sage, Hiram Sibley, and one graduate of the Mechanic Arts Department.[30] For the fourth slot, the trustees passed over the Halsey

27 Morris to White, February 27, 1883, in Sibley College Papers.

28 May 19, 1883, p. 8.

29 Coon, Falkenau, and Halsey, Petition to Board of Trustees, *ca.* 1884, in Trustees' Papers.

30 Special Committee Minutes, Trustees' Papers; Cornell University, Board of Trustees, Minutes, June 18, 1884, Cornell University Archives, Ithaca, N.Y. (hereafter cited as Trustees' Minutes).

group's recommendation of John Saylor Coon and picked Walter Craig Kerr, successful consulting engineer with the Westinghouse Machine Company in Pittsburgh. He was not, nor was there on the committee, an engineering educator.[31]

Kerr's appointment probably came partly because of a long and detailed letter he wrote to Henry W. Sage in 1884, suggesting improvements in Sibley College which would be possible with the recent gift which Hiram Sibley had made to the school. Morris' conception of the field was too narrow and limited, in Kerr's opinion, and what was needed was the "personal effort and immediate management of one of the several prominent Mechanical Engineers of the day."[32]

From the very first there were apparently only two persons seriously considered by the Sibley College Revision Committee, William Kent and Robert Thurston. Thurston's background has been discussed. Kent had been Thurston's assistant, a steel engineer, and a boiler expert, and was some years younger. White, and probably the others, were inclined toward Thurston but feared he would not leave his Stevens post. Each was invited to the campus and given a grand tour, then asked to submit in writing his recommendations for improving Sibley College and making it truly a school of mechanical engineering. Kent and Thurston were friends professionally and socially and knew what was going on. Nor was Andrew D. White unacquainted with Thurston's social and intellectual background. White's wife was the daughter of the same Edward McGill who had started young Robert Thurston on his route away from the shop some twenty years before. The Whites and the Thurstons had immediate social rapport.[33]

Kent's and Thurston's recommendations were quite similar in scope and content and represented in part a response to nearly twenty years of criticism from the elite of mechanical engineering. One major point expressed by both was that Sibley College should be entirely in charge of one chief professor,

31 Undated report of Revision Committee, June 17, 1888, file, Trustees' Papers. For further information on Kerr, see Albert W. Smith, *A Biography of Walter Craig Kerr* (New York: American Society of Mechanical Engineers, 1927).

32 June 14, 1884, in Trustees' Papers.

33 Durand, *Thurston*, pp. 96–105.

which would enable it to be "organized like an Industrial Establishment with one general manager" with the necessary authority to carry out all his ideas "in all the subordinate departments of the College."[34]

The committee of four took these suggestions seriously, recommended the appointment of a professor and director with "large powers" at a then phenomenal $4,500 yearly salary. They also recommended Thurston for the job, accepting his conditions in June, 1885. These conditions related to his status as "Director" of Sibley College and spelled out in great detail the relationships between the director and the faculty, administration, and trustees in terms of duties and responsibilities as well as rights and privileges. His proposal is a model of bureaucratic systematization.[35]

Thurston went to work at once to raise the entrance requirements in order to eliminate much of the elementary work in the first two years of the curriculum. The curriculum itself was reorganized to include basic science, higher mathematics, and language in the first two years, with courses in steam engineering and other specialized subjects in the last two years, with some shop experience and training in drafting throughout the program. His aim was to perfect his model of the pure technical school, which he had begun at Stevens, and, as he wrote to Andrew D. White in December of 1885, "the more rapidly the lower portion of the work is forced back into the preparatory and lower schools, and the more the colleges reach upwards into the higher fields of investigation, the more they will accomplish for the world."[36]

The new director of Sibley College was able to get what he wanted from the trustees because of his ability in communicating with and influencing them. In contrast to John Sweet, who seems to have communicated with the Executive Committee about two or three times in his seven years at Sibley, Thurston

---

34 Kent's recommendations are included in a letter to Henry W. Sage, April 22, 1885, in Trustees' Papers.

35 Undated report of Revision Committee, June 17, 1888, file, in Trustees' Papers. Thurston's recommendations are given in a letter to White, May 26, 1885, quoted in Durand, *Thurston*, pp. 103–5.

36 December 20, 1885, in Andrew Dickson White Papers, Cornell University Archives, Ithaca, N.Y. (hereafter cited as White Papers).

wrote letters to and spoke before the Committee on practically every occasion they met. He also communicated extensively with Andrew D. White (who retired shortly after Thurston accepted the Sibley position but who nevertheless remained influential in university affairs) and with the university presidents and chairmen of the Board of Trustees. He hammered at them constantly for the completion of his original plan of organization.[37]

One of the keys to his program was the increase in both size and quality of the faculty of the College, which grew in his eighteen years there from seven to forty-three and included highly qualified men.[38] Many of the men he recommended for positions—like William F. Durand of the Naval Engineer Corps, Harvey Williams, a designer with Yale and Towne, and H. Wade Hibbard, first a locomotive designer and later an engineering professor at the University of Minnesota—were experienced in or at least knowledgeable of the shop world.[39] Thurston was clever enough not to alienate the shop culture elite and thus lose their vital support. He attracted such men by strongly urging the trustees to raise faculty salaries, such as in 1890, when he wrote the Executive Committee that the newly acquired land-grant funds should be "very largely, if not mainly, devoted to placing the mechanic arts department in a position of security against competition and insuring their permanent prosperity." The school of agriculture, he added, did not need the money, since the current trend was "toward the technical."[40] If he was forced to hire "day laborers" as instructors, he wrote Andrew D. White in 1893, he could not achieve the efficiency of which the school was capable.[41]

One relatively easy way Thurston found to increase the

37 For example, see Thurston to President and Executive Committee, February 10, 1890, in Ex. Com. Papers.

38 Durand, *Thurston*, p. 110.

39 Thurston to Executive Committee, September 1, 1888, August 20, 1891, and November 28, 1897, in Ex. Com. Papers.

40 Thurston to Executive Committee, October 1, 1890, in Ex. Com. Papers. Thurston sent an appeal to the ASME concerning the same subject, suggesting a committee on technical education which would lobby for more land-grant money for technical departments and less for agriculture and "less needed specialties." Notation of letter from Thurston to the Council of the ASME in Minutes of the Council of the American Society of Mechanical Engineers (hereafter cited as ASME Council Minutes), in the office of the Secretary, American Society of Mechanical Engineers, New York, N.Y., September 7, 1887.

41 May 22, 1893, in White Papers.

quality of instruction was to implement Walter Kerr's idea of nonresident or guest lecturers. Thurston was able, through his personal prestige, to secure such speakers as Francis A. Walker, Alexander Graham Bell, Samuel P. Langley, Edward Atkinson, Charles T. Porter, E. D. Leavitt, Frank J. Sprague, Elihu Thompson, J. F. Holloway, Eckley B. Coxe, and dozens of other successful engineers and manufacturers.[42] One stellar attraction was Andrew Carnegie, who accepted an invitation in 1888 with the following words: "I am neither mechanic nor engineer, nor am I scientific. The fact is I don't amount to anything in any industrial department. I seem to have had a knack of utilizing those that do know better than myself." Carnegie's nephew was enrolled at Cornell, and the friendship between Carnegie and Thurston was a strong one.[43]

Thurston maintained good relationships with much of the manufacturing and mechanical engineering world. He corresponded with E. D. Leavitt, a dean of shop culture, with F. R. Hutton, secretary of the ASME, with inventors, engineers, and manufacturers like George Westinghouse, Hiram Maxim, and John Ericsson. He persuaded some of these men to give machinery and testing equipment to the college, others he induced to speak before his students, and others were potential employers for his students.[44] Elihu Thompson, the electrical inventor, engineer, and manufacturer, wrote to Thurston in regard to Thompson's difficulty in finding the right men for the various departments in his firm.[45] Thurston frequently brought potential employers to Ithaca to see his college and shops. Cyrus McCormick, after a visit in 1894, wrote Thurston of his approval of his methods and equipment.[46] Thurston also kept up his contacts with former and present naval engineers like Engineer-in-Chief George Melville and consulting engineer Charles W. Copeland.[47]

[42] See correspondence between Thurston and these individuals, 1885–88, in Thurston Papers.

[43] October 26, 1888, in Thurston Papers.

[44] Hutton to Thurston, March 27, 1890; Leavitt to Thurston, September 20, 1889; Westinghouse to Thurston, October 17, 1887; Maxim to Thurston, September 6, 1894; Ericsson to Thurston, January 5, 1887; all in Thurston Papers.

[45] Thompson to Thurston, July 1, 1888, in Thurston Papers.

[46] September 6, 1894, in Thurston Papers.

[47] See especially Melville to Thurston, June 23, 1892, and Copeland to Thurston, November 14, 1888, in Thurston Papers.

Making of air compressors, steam pumps, and engines at the Norwalk Iron
Works, South Norwalk, Conn., 1880 (*Scientific American*).

Extensive research into the methods of engineering instruction was only one reason why Thurston kept up his wide acquaintance with British, German, and French engineers and educators. Not only did he want to know everything that was going on in engineering education, but he also kept up on the latest advances in engineering science through correspondence with such European scientists as Lord Kelvin.[48] As he wrote to A. D. White in 1885, "I have more or less correspondence in every great school in Europe."[49]

Publicity about the college and the new program was a key feature of Thurston's development plan. He wrote White soon after he had taken over his new job that he had published a long article in the *Scientific American* on Sibley and had ordered 8,500 reprints for promotional purposes, more specifically for distribution to all the members of the ASCE, AIME, and ASME![50] Thurston also ran frequent reports on Sibley and successes of its graduates in the *American Machinist*, the *Scientific Machinist*, and similar engineering periodicals.[51]

Another method Thurston used to boost both his students and the program was the practice of organizing several-week field trips by parties of students to industrial plants and engineering shops around the Northeast. Many practicing engineers were glad to assist Thurston in this plan. One, John Fritz of the Bethlehem Steel Company, assured Thurston that he would "make such arrangements as will enable them to see the works and will have proper persons go with them to explain anything they may want to know."[52] About the only refusal of any kind in all the correspondence in Thurston's papers about the field trip program was the refusal in 1893 by Chauncey M. DePew to run a special train over the Lake Shore Railroad for the students.[53] Less formal was Thurston's sending four young students to visit the world-famous engineer John Ericsson in

48 For example, Kelvin to Thurston, August 2, 1887, in Thurston Papers.
49 Thurston to White, October 26, 1885, in White Papers.
50 *Ibid.*
51 For example, see "Cornell University," *Amer. Mach.*, December 20, 1888, p. 4; S. G. Pollard, "Notes on the Mechanical and Electrical Engineering Departments, Cornell University," *Amer. Mach.*, January 22, 1891, p. 12.
52 Fritz to Thurston, March 17, 1887, in Thurston Papers.
53 DePew to Thurston, March 10, 1893, in Thurston Papers.

1887.[54] He was not shy about letting the Executive Committee of the Board of Trustees of the university know when his students made good in the mechanical engineering elite. He reported to them in 1890 that four Cornell men had presented papers at the last meeting of the ASME.[55]

By 1890, only five years after he went to Cornell, Thurston could write to the president and the Executive Committee that his work was now done, "practically, so far as Sibley College is concerned, considered as a school of elementary instruction in professional matters simply. If anything more is to be done, it must be in the inauguration of the higher work of advanced and specialized schools."[56] He proudly wrote to Andrew D. White in 1893 that his conception of the purely professional school with no attempt at general education was highly efficient and successful. "Our men are better educated and better trained, professionally, than when we were trying to give them a hodge-podge, neither education nor professional training."[57] Perhaps he was mistaking the source of his success. What had probably turned the eyes of the engineering world on the Cornell experiment was the brilliant way in which Thurston had been able to make the professional school replace the important functions of shop culture in such areas as socialization and orientation. If much of what Thurston accomplished along these lines was by dint of his own personal prestige, power, and acquaintance, he nevertheless created institutional frameworks within which the relationship between school and shop could be expanded and enriched.

The master of Sibley College did have problems and critics, however. For example, he was not very successful in establishing at Cornell one of his pet projects, an industrial and engineering testing laboratory. He was also troubled by the problem of too many students applying for admission to Sibley and by feelings on the part of some critics, both within the university

54 Ericsson to Thurston, March 16, 1887, in Thurston Papers.
55 November 18, 1890, in Ex. Com. Papers.
56 October 18, 1890, in Ex. Com. Papers.
57 January 20, 1893, in White Papers. For a description of life as an engineering student at Cornell in the eighties, by a man who then worked for John Sweet and later for George Corliss, see Embury A. Hitchcock, *My 50 Years in Engineering* (Caldwell, Idaho: Caxton Printers, Ltd., 1939), pp. 31–45.

and without, that Cornell and particularly Sibley College was not fulfilling its obligation under the land-grant act or Ezra Cornell's bequest. Cornell was not, so went the argument, educating and providing opportunity for the sons and daughters of poor but honest citizens, artisans, and farmers. Thurston was among the first to recognize the inconsistency of ideology and practice and wrote his friend Andrew D. White that "a difficulty arises here from the evident and indisputable intention of the founders of the University to make it useful to the masses." Thurston would, he wrote, prefer to have quality and not quantity.[58]

He worried about the rapidly increasing numbers of applicants for places in the engineering schools and about the students who, once in school, couldn't make it. For the latter, Thurston proposed a "mechanic arts" course which would exclude higher mathematics and the experimental laboratory work. This, he wrote ex-President White, would allow the possibility for distinction for "good mechanics, and honest workers, and good students too, unequal, however, to the handling of the higher mathematics."[59] Actually, Thurston's actions in the following years led away from any sort of special course for mechanics. On several occasions he denied charges in the technical journals that Cornell did not allow credit for shop experience to stand against a lack of academic entrance requirements.[60] In truth, the standards became more and more those possessed by the high school graduate and it was the rare boy from the shops who could get in.

Thurston's trend toward the pure professional school led him to lack sympathy for the aims and ideas of old Hiram Sibley, the school's continuing benefactor. He wrote to White in 1886:

It is my impression . . . that Mr. Sibley's ideas are not keeping pace with the growth of this work, and that he has little idea of more than a workshop in which apprentices may be taught the old fashioned ways of "making

---

58 December 20, 1885, in White Papers.

59 Thurston to President and Executive Committee, September 18, 1889, in Ex. Com. Papers; Thurston to White, December 20, 1885, in White Papers.

60 For example, see Robert H. Thurston, Letter to the Editor, "Entrance Requirements, Endowments and Scholarships at Cornell University," *Amer. Mach.*, March 5, 1886, p. 267.

things." I do not think that he realizes what enormous advances have been made in the Mechanic Arts since he was a boy and learned the art of making a shoe of the old fashioned kind, by looking on for an hour, and cannot perceive of the precision of modern ways and methods.[61]

Thurston concluded that Sibley was proud of the college and could be educated. As soon as Thurston made it clear, however, that his program would harmonize head and hand and turn out bold, creative engineers, Sibley gave additional, large sums of money for the college's support.

Above all, however, Robert H. Thurston was an engineering educator, even though, like many educators, he was also accepted in the mechanical engineering elite. His ideas of what technical education should be had fully evolved by 1893, when he presented a nearly two-hundred-page paper to the ASME on the subject of technical education in the United States. It was a complete system of education for mechanical engineering, and its presentation coincided with the formation of the Society for the Promotion of Engineering Education and the professional emergence of the engineering educator as a distinct type.[62]

The experience of Cornell is a study in the types of conflict which went on in schools across the country. Within the walls of Sibley College, shop culture put up a losing battle against school culture. School culture, though it was the victor almost everywhere, was forced by the conflict to pay lip service to and frequently adjust its programs to the demands of shop culture and the mechanical engineering elite. That the educators were not forced to yield even further can be attributed to changes in the role of the mechanical engineer which took place during the developmental years of engineering education.

The men who led the innovative and scientific forces in the practice of mechanical engineering frequently disagreed with the aim of the technical school to replace the shop culture as an institution for socializing, educating, and professionalizing the mechanical engineer. They often spoke with disdain of

61 February 28, 1886, in White Papers.

62 "Technical Education in the United States," pp. 855–1013. The content of this paper is an elaboration of the structure and curriculum of his program at Sibley College already discussed.

"professional professors"—who have no natural or acquired taste for an appreciation of machinery or of machine construction. They have attained their positions simply because of what may be called general mental capacity, and would have done about as well in any other work of life they might have happened to choose. . . . As one engineer has said, they look down on manual training much as the preceptress in a young ladies' seminary might look upon instruction in laundry work for her pupils. She might have them take such instruction, not to make washerwomen of them, but that they might know how it should be done and criticise when necessary with sufficient intelligence to avoid being laughed at.[63]

The defenders of shop culture envisioned the mechanical engineer as a bold, creative craftsman, who could use higher mathematics and other tools of engineering science, but who was not narrowed by them. They thought of an elite corps of engineers when educators were contemplating and planning for a mass market for their competent but standardized products.

Those engineering educators who were most successful in breaking down the resistance of the shop culture defenders were those who, like Robert H. Thurston, were able to replace both the manifest functions of the shop, such as manual training, and also the latent functions, such as the socialization of the young professional, the making of contacts within the elite, and the inculcation of professional standards. The less successful attempts at replacing shop culture with school culture were those which tried to implement only the manual training functions and stressed production of articles for profit in the school shop, and those which failed to make any concession at all to the latent functions of shop culture and thus failed to prepare their graduates for the adjustment to the role of mechanical engineer in the machine shop.[64]

[63] Editorial on "Education of Machinists, Foremen and Engineers," *Amer. Mach.*, January 4, 1900, p. 1.

[64] The sheer growth of engineering education made it inevitable that eventually college graduates would dominate mechanical engineering. In 1870 one in eight or nine engineers was a college graduate; by 1916 half were engineering graduates. Mann, *Engineering Education*, p. 19.

American Engineer began its life as a technical journal in 1860 in New York City. It seems to have appealed to both mechanical and stationary engineers, and its early issues contain the first example of a significant number of individuals advertising themselves as mechanical engineers. John C. Merriam, the editor, was one of those who offered his services as a "mechanical engineer." The masthead announced that the magazine was "devoted to the interests of locomotive, marine, and stationary engineers." The weekly fare was not rich, but did include references to the work of Zerah Colburn, an article on the use of steam expansively, and another on the proper shape of cutting tools.[1]

By August of 1860 the journal was reporting the activities of an American Engineers' Association, which had recently been formed in New York, partly as a project of the editor, John C. Merriam. He reported in November the association's call to "join your brothers for the good of the profession."[2] The major concern which bound the members together was interest in the problems of the use and development of the steam engine in all its forms. By the end of November, the association was reputed to have sixty members and a committee on science and new inventions. A study of the twenty-one men who were elected as officers, committee chairmen, and members in December, 1860, reveals that none of these individuals was then or later became well known in mechanical engineering circles. There is no evidence that any of them, with the possible exception of Merriam himself, ever was a practicing mechanical engineer. The membership and the leadership of the organization was most likely composed of stationary or operating engineers.[3]

Some of the members and leaders, led by Merriam, wished

---

[1] For example, see *American Engineer* (New York) (hereafter cited as *Amer. Engr.* [N.Y.]), August 4, 1860, pp. 1, 8; *Amer. Engr.* (N.Y.), January 24, 1861, pp. 1–3.

[2] "The American Engineers' Association," *Amer. Engr.* (N.Y), November 1, 1860, p. 4.

[3] *Amer. Engr.* (N.Y.), November 22, 1860, p. 4; *Amer. Engr.* (N.Y.), December 6, 1860, p. 5.

to elevate the membership and prestige of the organization by inviting well-known mechanical engineers to join, thus making it into a professional association rather than a license-law lobby for engine drivers. Their effort to elect Benjamin F. Isherwood (prominent naval mechanical engineer) to membership brought no attendance from him, merely a letter of thanks for the honor. A further effort was made to place the names of several members of the mechanical engineering elite on the official ballot for new officers in March, 1861. Isherwood, A. L. Holley, Merriam, and Charles H. Haswell were nominated to fill the association's three vice-presidential slots.[4] Of the four, only Haswell was elected. Two others, elected on a write-in campaign, are unknown. Merriam complained on March 23 in his journal that unfortunately an insignificant cabal overthrew the regular slate and elected their own men. The results of the election showed a shift to a new group of leaders.[5]

The issue was whether the association would move in the direction of a professional mechanical engineering organization with emphasis on the presentation of papers on engineering problems, or whether it would become a lobby group devoted to securing effective license laws to protect competent stationary engineers. The latter group won, but neither the association nor the journal was heard from afterward. Twenty years later, Lycurgus B. Moore would tell the members of the American Society of Mechanical Engineers that "belittling influences" had "heretofore nullified all previous attempts to found a national association for mechanical engineers," which suggests that there were many such attempts, that the American Engineers' Association was not unique, and also that its organizational history was typical of the experience of such groups.[6]

Since the mechanical engineer had emerged as a type by 1855, one wonders why a lasting and influential professional association was not formed until 1880. The period from 1851 to 1876 saw the United States leading the world in technical innovation and precision in the areas of mechanical engineering employment and activity, such as in machine tools, steam

4 "American Engineers' Association," *Amer. Engr.* (N.Y.), March 9, 1861, p. 4.
5 "American Engineers' Association," *Amer. Engr.* (N.Y.), March 23, 1861, p. 4.
6 "Proceedings of Meeting," ASME, *Trans.*, II (1881), 410.

engines, and pumps. The American Society of Civil Engineers was formed in 1852 but languished from 1855 to 1867, then flourished in the late sixties and became a strong organization. The American Institute of Mining Engineers, partly an offshoot of the ASCE, held its first meeting in 1871 and thereafter grew and was active.

When one considers that serious professional activity in the engineering occupations did not begin until the late 1860's and early 1870's, it does not seem so unusual that the mechanical engineers did not organize until 1880. The answer may be quite simple: before 1870 there may have been no conscious need for such professional activity. Many dozens of occupational specialties which had existed before the 1870's did not begin to develop professional associations until the two or three decades following the Civil War. Individuals having occupational or other ties in common sought in this period to preserve or increase their influence and esteem in society by forming collective associations. This was true of both professional and nonprofessional segments of society. In consequence the formation of the ASME was not a unique act, but rather a part of the rapid trend toward a more complex organizational structure of society.

Some mechanical engineers, it is true, did find temporary homes in the ASCE and later in the AIME, where they were able to present papers occasionally and to find a few compatriots. Before 1885, the transactions of both of these organizations contain papers on mechanical engineering subjects delivered by men practicing in the field. Though some mechanical engineers retained their membership in the other two organizations after joining the ASME in the 1880's, none wished to or was allowed to publish papers in their journals. If the slowness of the ASME in appearing seems surprising, its early years are remarkable for the successful management and organization the group displayed. This is in marked contrast to the ASCE and the AIME, both of which had rather difficult times in their first years. However, the ASME was founded nearly ten years later (a very significant ten years) and had the model of the successes and the mistakes of the other two before it.

Mechanical engineers had been prominent participants in

international exhibitions since the 1860's, and these affairs, often held in foreign engineering capitals, provided a meeting place for them. This was particularly true of the Centennial Exhibition in Philadelphia in 1876. It was the site of the engineers' conference on education and by its very nature made the different branches of engineering self-conscious of their separate existence and of the common bonds between members of specialties. Machinery Hall at Philadelphia was, as it had been at previous exhibitions, the focal point of interest for mechanical engineers. These shows provided a formal way of meeting and exchanging information and ideas which enriched and enlarged shop culture.[7]

Another product of the 1870's which led to the growth and elaboration of shop culture were the technical periodicals devoted particularly to mechanical engineering and its problems. Probably the most significant of all of these periodicals was the *American Machinist*, founded in New York in 1877, just a year after the Centennial. Jackson Bailey, its editor, began in late 1879 a correspondence with John Edson Sweet, dean of shop culture, concerning the establishment of a national society to be devoted to the advancement of mechanical engineering. Sweet was interested but lacked the organizational drive necessary to get the project under way and was reluctant (from personal modesty, we are are told) to issue a call in his own name. Bailey visited him in Syracuse, where he had gone after his Cornell experience to found the Straight Line Engine Company, and persuaded Sweet to prepare a list of persons to whom invitations should be sent. Sweet contacted one of the best-known members of the mechanical engineering elite, Alexander Lyman Holley, and the leading mechanical engineering educator, Robert H. Thurston. Jointly they issued a letter calling for an organizational meeting at the office of the *American Machinist* on the sixteenth of February, 1880.[8]

Key feature of the conference in the *American Machinist* offices was the address by Alexander Lyman Holley on "The

---

7 Monte A. Calvert, "American Technology at World Fairs, 1851–1876" (unpublished Master's thesis, University of Delaware, 1962).

8 Hutton, *History*, pp. 4–5.

Field of Mechanical Engineering." He pointed out how much mechanical engineering underlaid the structure and uses of civil engineering. Replying in advance to possible criticisms that the former was not worthy of professional status because it involved only the assembling of parts to make machines, Holley said: "Should the architect and the civil engineer say that the mere molding and assembling of members is not worthy of a professional name and *status*, the mechanical engineer may reply that the mere calculation of strains from known formulae, and the mere groupings of conventional forms, is no more worthy." He defined the aims of the proposed organization as: the collection and diffusion of knowledge, the advantages of personal acquaintance among members, the value of writing and discussing papers, and the significance of the endorsement of a high quality of membership.[9]

The preliminary conference decided to found an organization and made plans for a meeting to be held April 7. This second meeting was also a personal affair; the guest list was compiled by three men and consisted of the first list plus names recommended by the men on that list. At no time in these early negotiations was there any attempt to systematically contact every practicing mechanical engineer in the United States, nor was there any effort to isolate large groups who might be contacted, such as all graduates of technical schools holding the M.E. or B.M.E. degree. The ASME was, like the shop culture it was to supplement, a personal affair, almost a kind of gentlemen's club.[10]

The editors of the technical journals greeted the incipient association with considerable approval. *American Engineer* of Chicago expressed surprise that "it was not inaugurated years ago."[11] *Iron Age* noted that until now papers had been presented to other societies, "where naturally they did not elicit the discussion which they merited, nor did they go directly into the hands of those whom they were chiefly designed to reach."[12]

The journals were not without caution, however. *American*

9 ASME, *Trans.*, I (1880), 1–3.
10 Hutton, *History*, p. 15.
11 I (March, 1880), 40.
12 Quoted in *Amer. Mach.*, March 13, 1880, p. 8.

Machinist, whose editor Jackson Bailey had been the originator of the idea, warned in January, 1880, that the plan for the society must be "broad and liberal, and not restricted to engineers or experts, but open to every toiler who works with his hands." The journal offered the example of the Syracuse, New York, society, founded through the efforts of John Sweet, in which mechanic and engineer mingled with profit.[13] Fearing perhaps a criterion for membership based on consulting practice, Bailey warned that the new organization must also include those "who as foremen, designers, superintendents, or proprietors of works direct operations requiring the exercise of engineering skill."[14] By April 3, 1880, the work of the various committees was well known and Jackson Bailey disapproved via his editorial columns. He was aware, he wrote, that the membership qualifications being considered were so restrictive that they would exclude many important engineers. "No reader of the *American Machinist* needs to be told that there are engineers, contributors to the technical press, practical men in the shops, and others, who are writing their names in letters of steel upon the mechanical records of the time, the results of whose works will live when some of the book-writers (and compilers) are forgotten."[15] Bailey and other technical journalists feared that rigid, standardized qualifications for membership would be enacted, but this fear was unfounded, as were Bailey's hopes that every machinist from the shop would be eligible for membership. The American Society of Mechanical Engineers was and would remain for at least the first twenty years of its life an elite organization of the profession and the shop, into which one entered through the personal recommendation so characteristic of shop culture.

The qualifications adopted by the ASME divided membership into three categories: full member, associate member, and juniors. Members could be mechanical, civil, mining, metallurgi-

---

13 "Mechanical Engineers' Association," *Amer. Mach.,* January 3, 1880, p. 8.

14 *Amer. Mach.,* February 28, 1880, p. 8. As late as 1907 a past president of the ASME wrote the secretary a strong plea to avoid determining membership on the basis of present employment. Copy of letter, C. W. Hunt to Calvin W. Rice, January 24, 1907, in Taylor Collection.

15 *Amer. Mach.,* April 3, 1880, p. 8.

cal, and naval engineers and architects who were considered "in the opinion of the Council, competent to take charge of work in his department, either as a designer or constructor, or else he must have been connected with same as a teacher." Associates included those who had "such a knowledge of or connection with applied science as qualifie[d] him . . . to co-operate with engineers in the advancement of professional knowledge." This second category was one which enabled the council of the society, with the approval of five members, to admit those engineer-entrepreneurs they desired, without open-ing the gates to any and all entrepreneurs. Juniors was a cate-gory for precocious young men whose age and experience did not justify their admittance as full members.[16]

This informal membership criterion was intentional; in public ideological statements the mechanical engineers themselves averred that its purpose was to avoid "the forcing of a Pro-crustean uniformity of training and experience." As Frederick R. Hutton wrote in the official history of the ASME in 1914: "The policy of broad interpretation of the eligibility require-ment has been one of the cornerstones of the success of the Society."[17] The need was not felt even for a standing member-ship committee until 1904. Obviously this informality was a doubled-edged sword and could be used to keep out, on per-sonal grounds, those who were well-qualified engineers, but it was rarely used for this purpose.[18] Instead it seems to have been insurance that the regularization of previous training and experi-ence which the technical schools were setting up would not be repeated in the national professional association. It assured that the ASME would be a stronghold of shop culture and its defense.

Not everyone was happy with the *apparent* looseness of the standards and qualifications for membership, and several at-tempts were made over the first quarter-century of the society's existence to raise both age and experience qualifications for

---

16 "Rules of the American Society of Mechanical Engineers," ASME, *Trans.*, II (1881), vii.

17 Hutton, *History*, pp. 17–18.

18 ASME Council Minutes, August 8, 1883, November 10–11, 1885, and March 24, 1886, show how the organization dealt with such problems.

the various grades of membership. In 1894 such sometime exponents of shop culture as F. R. Hutton and Oberlin Smith led one such movement. Significantly, they wished to require at least five years' teaching experience for the category of teachers of engineering and to make the requirements for associate and junior members less precise.[19] The analogy with the shop culture idea of easy entry but high internal standards for advancement is obvious.

Those who attended the first organizational meeting in February of 1880 were all prominent men in shop culture. No mechanical engineering educators were present. By the April 7 organizational meeting, several mechanical engineering educators responded to the call, but the predominance of shop men was still pronounced. The members listed in the first catalogue of the ASME, published in September, 1880, covered a wide spectrum of occupational roles, but nearly every man was relatively well known within the engineering world.[20]

Predominant in numbers were the owners, superintendents, and chief engineers of shop installations. This group was tempered with a sprinkling of educators, professional technical journalists, consulting engineers, naval engineers in service, and others. This distribution of membership remained constant throughout the period under discussion, indicating that there was no great change in the occupational role of the members of the ASME, at least through about 1900 to 1905. Between 1880, when the membership stood at a little over a hundred, and 1907, when the total United States membership was 2,957, the society underwent a steady, continuous growth. Throughout this period of growth the membership clearly in the shop culture tradition ran to at least 50 per cent of the total. A survey of the society made in 1907 indicated (of a total of 2,957), unclassified, 10.3 per cent; army, navy, and marine, 1 per cent; hydraulic, 0.4 per cent; patent attorneys, 0.8 per cent; journalists, 1 per cent; mining and metallurgy, 1 per cent; engineering contractors, 1.6 per cent; testing and inspecting, 1.6 per cent;

19 "Proceedings of the New York Meeting," ASME, *Trans.*, XVI (1894–95), 3–39.

20 ASME, *First Catalogue of the American Society of Mechanical Engineers* (New York: American Society of Mechanical Engineers, 1880).

operating engineer, 1.8 per cent; locomotive and railway, 1.9 per cent; electrical, 2.2 per cent; professors and teachers, 6.3 per cent; office practitioners, 16.5 per cent. The remaining 53.6 per cent of the membership was entrepreneurial and shop-centered, breaking down into draftsman and designer, 4 per cent; local manager, 5.2 per cent; shop executive, 11.8 per cent; and manufacturer, 32.6 per cent. Even within the shop category, those who were clearly businessmen, as much as they were engineers, dominated the membership numerically. The figures for common occupations of the young college graduate (draftsman) indicate that large numbers of technical school graduates were not joining the ASME. Perhaps they were not invited to join. On the other hand, they may not have felt there was any advantage in joining or much chance for advancement to full membership without acquaintance in the shop culture aristocracy. As further evidence of this, the membership of the ASME grew from about one hundred in 1880 to about one thousand in 1890, a decade in which there were few technical graduates; however, from 1890 to 1900, a decade in which the number of technical graduates rose into the thousands, membership growth in the ASME actually leveled off in terms of rate and increased only another thousand to a total of just under two thousand members. The United States census figures do not begin to delineate mechanical engineers separately until 1910, but combined figures for mechanical and electrical engineering indicate that by 1900 at least ten thousand individuals listed themselves as mechanical engineers, and there were probably more than that. Representing, as it did, less than one quarter of the total practitioners in the profession (a good many of the manufacturers probably listed themselves as something other than mechanical engineer), the ASME was not a broadly based mass membership organization, but was, in spite of its numbers, an elite organization which offered little to the young technical graduate with no contacts in the shop culture aristocracy.[21]

---

[21] See graph on growth in number of members in Hutton, *History*, p. 76; see chart on number and percentage of members by occupation, F. R. Hutton, "The Mechanical Engineer and the Function of the Engineering Society," ASME, *Trans.*, XXIX (1907), 325–26.

Production of universal chucks at the E. Horton and Son Co.,
Windsor Locks, Conn., 1880 (*Scientific American*).

Mere membership, moreover, did not imply participation in the organization and attendance of meetings. At the two meetings held in 1890 when the membership was about 1,000, attendance was 164 (at the first) and 126 (at the second). This was a year when both meetings were not held in New York City. In 1900, when the membership stood at about 2,000, the Cincinnati meeting drew 144 and the New York City meeting drew 467. In 1900, then, one-quarter of the membership in the ASME had enough interest to attend the annual convention, and the membership represented about one-fifth of the total practitioners. Participating membership in the ASME was then, in the period from 1890 to 1910, less than a twentieth of the total number of practicing mechanical engineers in the United States. Participation in terms of votes cast on questions before the society and for officers ran to a slightly lower fraction.[22]

In spite of the fact that the ASME was not a mass membership or participation organization, fears were expressed by members that it was becoming so.[23] The jump from meetings attended by perhaps a hundred men in the 1880's to about five hundred by 1900 was not inconsiderable and threatened the principle that everyone knew everyone else by sight. This change is reflected in the *Transactions* of the society, in which, before December, 1902, every person who spoke in a meeting was identified by name, full name if he was not too well known and last name if there could be no question of who he was. In reports of the transactions of the December, 1902, meeting, speakers in a discussion were referred to for the first time as "A Member." The practice increased in the years to follow. There was at the same meeting a vigorous discussion of junior membership and how applicants for that status should be screened, in view of the growth and increasing impersonality of the society.[24] In a 1908 discussion, hearsay information in regard to membership applications was attacked and the suggestion made that it was becoming increasingly difficult for the

---

[22] Number of attendants at each meeting, Hutton, *History*, pp. 196–97.
[23] On the growing impersonality of the ASME see Colvin, *Sixty Years*, p. 69, which cites the intimacy of the early years. There was about these meetings "the air of the small-town college alumni gathering and the old-home-week reunion."
[24] "Proceedings of the New York Meeting," ASME, *Trans.*, XXIV (1902–3), 3–77.

old system of recommendation to function efficiently.[25] From the mid-nineties on, increasing concern was shown by members that many of the young men came to the meetings and left without meeting or talking to anyone. Special committees were set up to hand out name tags and to make efforts to introduce the younger men around.

At least part of the lack of participation and lack of interest in membership in the ASME on the part of many practicing mechanical engineers can be traced to the geographical bias of membership and to the existence of an eastern establishment. In 1903, of 2,577 members of the ASME, 802 lived in New York, 377 in Pennsylvania, 249 in Massachusetts, 188 in Ohio, 164 in Illinois, 130 in New Jersey, and 111 in Connecticut. The rest of the membership was scattered among many states, with a surprising number having no more than could be counted on one's fingers. The three Middle Atlantic States of New York, Pennsylvania, and New Jersey, combined with Massachusetts and Connecticut (lower New England), accounted for 1,669 of the 2,577 total, or about two-thirds. With the exception of the Chicago and Ohio enclaves, no other major aggregations of members existed.[26] One could make a case for the idea that this was simply where the mechanical engineers of the country were located. By 1903, however, the Midwestern engineering schools like Purdue, Wisconsin, and Minnesota were turning out large numbers of graduates and not all of them went East to work.[27]

That some of the leading ASME members felt that the country ended at Pennsylvania's western border is indicated by the remark of Acting Secretary Lycurgus B. Moore, reporting the February 16, 1880, organizational meeting, that "about forty

25 "The Annual Reports of the Council and Committees, 1908," ASME, *Trans.*, XXX (1908), 551. By 1887 the ASME Council had to take action on the problem of increasingly lukewarm recommendations to membership. ASME Council Minutes, October 20, 1887. By 1890 the list of candidates had grown so large that it was impossible for the whole council to consider it; an *ad hoc* committee was set up to screen applicants. ASME Council Minutes, April 15, 1890.

26 Geographical summary of members by state or country, ASME, *Trans.*, XXV (1903-4), v-vi.

27 Engineers working in the West and Midwest often had trouble getting to know enough ASME members personally to prepare a complete application for membership. One letter relating to the situation in Louisville was read before the council. ASME Council Minutes, October 13, 1882.

gentlemen attended that meeting, some coming from as far west as Ohio."[28] The headquarters of the society was, of course, in New York City, where more mechanical engineers practiced than in any other city (more than in most states!) and where perhaps a fifth of the total ASME membership lived. One of the two yearly meetings was held in New York, the other, to stimulate and encourage "non-resident members" as they were sometimes called, was held in other parts of the country, but always in industrial cities, such as Hartford, Altoona, Cleveland, Pittsburgh, Chicago, Boston, and Philadelphia. As the expenses of running the headquarters building or "house" of the society increased over the years, members outside of a one-hundred-mile radius of New York City frequently complained that they were being cheated, particularly on several occasions when attempts were made to raise the dues. An editorial in the *American Machinist* in 1901 described with some feeling the plight of "a member who lives a long distance from New York and seldom attends a meeting and who has no vote except for new members and for a list of officers whose nomination is equivalent to election, inevitably feels that he is shut out by circumstances from anything like a full participation in the affairs of the society and he unavoidably loses interest in it."[29] A reader wrote to the journal, addressing the problem more directly, saying, "It looks very much as if the entire membership were being taxed to maintain a club-house and library for the members in and around New York."[30]

Numerous suggestions were made for ways to correct this situation. In 1888 at the Scranton, Pennsylvania, meeting, it was suggested that either the society should have one meeting per year, but meet in sections in a number of major cities, or try the less drastic expedient of adding a third meeting each year outside New York City. The motion was defeated when George Babcock, a leading machine shop owner, and Robert H. Thurs-

---

28 "Acting Secretary's Report," ASME, *Trans.*, I (1880), 3.

29 "The American Society of Mechanical Engineers, Meeting and the Vote," *Amer. Mach.*, December 12, 1901, p. 1365.

30 November 28, 1901, p. 1302. Even members as close as Philadelphia sometimes dragged their feet regarding appropriations for expanding the society's house operation in New York City. Frederick W. Taylor to Morris L. Cooke, April 29, 1907, in Taylor Collection.

ton opposed the measures openly. Babcock's and Thurston's speeches against the change drew great applause, something that rarely happened at meetings.[31] Indeed, the trend seemed to be toward greater centralization of the power structure of the society. In 1905 ASME President Frederick W. Taylor posed the question of whether it was better to fill vacancies in standing committees in such a way as to achieve "wide geographical distribution" or to select those living "within a relatively short radius of the Society's headquarters." Taylor suggested, and the society at once approved, the idea that the most important consideration was the "efficient" transaction of the society's business and the ease of gathering the members of a committee on short notice.[32] William F. Durfee resigned as chairman of an ASME committee when he moved to Pennsylvania, and the council decided to pick a replacement "from eligible city members of the Society."[33]

It is clear that the ASME was an organization of, by, and for the shop culture elite. Secondly, the organization of the society was such that it precluded any sort of working democracy. At the time of the founding of the organization, the committee to form organization and rules was composed of four shop owners, two technical journalists, and one educator. The powerful nominating committee, however, was composed completely of representatives of shop culture, specifically A. L. Holley, John Sweet, E. D. Leavitt, Charles T. Porter, and Henry R. Worthington.[34] Holley was on both committees, and even the official historian of the ASME, F. R. Hutton, admitted in 1914 that "there seems little doubt that these rules were drafted by Mr. Holley, and sent for criticism to his colleagues, and found acceptable by them." There was some criticism at the founding that the society would be hurt by a predominance of manufacturers and businessmen among its prime movers, but this feeling was apparently dissipated rather easily.[35]

31 "Proceedings of the Scranton Meeting," ASME, *Trans.*, X (1888–89), 3–31.
32 "Proceedings of the New York Meeting," ASME, *Trans.*, XXVII (1905-6), 68–69.
33 ASME Council Minutes, September 7, 1887.
34 Hutton, *History*, p. 11.
35 *Ibid.*, pp. 18–19. See Holley's dominant role in running the ASME meeting of October 6, 1880, in ASME Council Minutes for that date.

The plan of government set up by the founders of the ASME was one which began with the nomination by a special committee of one candidate for president, three for vice-president, and three for managers of the society. This slate was then approved by the voting members of the society (about 20 to 25 per cent actually bothered to vote). These officers formed the council of the society which was the supreme policy-making and governing body. The executive officer of the society was the secretary, appointed yearly by the council (for the express purpose of removing that office from partisan politics). This secretaryship became a permanent job, and the ASME had only two secretaries between 1883 and 1934. The presidency was primarily an honorary position and represented little actual power. Clearly, all power was vested in the council. It was assumed that its members would be better qualified to choose the secretary because they "would be cautious and painstaking in choosing their executive to a degree which the irresponsible voter at large could neither recognize or live up to." It would have been nearly impossible for the formal leadership of the ASME to have been anything other than a self-perpetuating oligarchy, controlling as it did the selection of its own successors.[36]

There were, from time to time, slight murmurs of discontent among the members about this situation and some talk of popular election of officers. Interestingly enough, the leadership was disappointed that so few members took part in the voting, and they took various measures to try to increase the percentage of voting members.[37] The editors of *Mechanics* stated the problem and the solution in 1882, when they commented that the slate proposed by the nominating committee was undoubtedly a good one, but that the method of election was wrong. The absence of an organized campaign, which might make it possible for write-in candidates to compete, made nomination tantamount to election. The president, who appointed the

[36] Hutton, *History*, p. 18.

[37] *Ibid.*, p. 30; ASME, *Trans.*, IV (1882–83), 14–16. On at least one occasion the council considered putting up two names for each office but rejected the idea as "not conducive to the harmony & success of the Society." ASME Council Minutes, November 16, 1891.

nominating committee, could control the succeeding election. Thus, the affairs of the society got into the hands of a ring, and the only way out of this impasse was to allow free nominations from the floor.[38]

Several attempts at raising the dues of the society presented a rather curious example of internal political action. The dues were raised in 1891 after a post card ballot which had the approval of the technical press.[39] A different situation arose, however, when the society discovered in 1901 that it was, and had been for some time, in debt. Expenses were again rising faster than income (the society's growth had tapered off), and the council decided the answer was another dues increase. Instead of submitting a letter ballot as before, the intention was announced of putting the measure to a vote of the attending membership at the annual meeting in New York.[40]

Traditionally the defenders of democracy in the society, the technical journals and their editors engaged themselves in a campaign to defeat the measure by amassing hundreds of negative proxy ballots from members who did not usually attend meetings. Matthias N. Forney, editor of the Railroad Gazette, and Fred J. Miller of the American Machinist were the leaders of the movement to defeat the dues increase, and they were successful. One council member, Francis H. Boyer (ASME manager, 1899–1902, an engineer for a refrigeration plant), became, according to the American Machinist, "somewhat hysterical over what he seemed to imagine was an invasion by people who had no right to express an opinion and by members of the Society who had no right to vote upon its affairs."[41]

Much of the opposition to the increase was based on the notion that the increased expense of running the society came largely from the continuously expanding house and headquarters operation, and many suggested that only members resident in New York City should have their dues increased. The inefficiency of the society's operations was, furthermore, particularly embarrassing since engineers (of all people) were

---

38 Mechanics, October 28, 1882, p. 273.
39 "Proceedings of the New York Meeting," ASME, Trans., XIII (1891–92), 35.
40 Hutton, History, pp. 210–11.
41 "The American Society of Mechanical Engineers, Meeting and the Vote," p. 1365.

supposed to be efficient. Many of the more wealthy engineers expressed the opinion that the dues should be increased if for no other reason than to make the organization a twenty-five-dollar affair, and one hoped that it might eventually become a hundred-dollar affair.[42]

Certainly there was much that was petty and insignificant in the minor squabbles within the ASME. This was so because the operations and meetings of the society were carefully controlled and differences over basic policy were very seldom made public. Partly, the apparent agreement was based on a broad consensus. Nevertheless, when a special committee on rules and methods was appointed to straighten out the society's affairs, the political structure was changed somewhat: a clause was inserted to "prevent vexatious discussions on past subjects"; the length of time required to pass constitutional amendments was specified as one year and that required to amend the by-laws was set at six months; and the council was given the power to amend the rules of the society at any meeting. If the ruling oligarchy had been challenged, it was certainly not ready to give in to demands for a more democratic running of the society's affairs.[43]

Cliques or interest groups did apparently exist, however, within the ruling oligarchy. F. R. Hutton, looking back on the affair of 1901 over the dues increase, described two factions, a progressive party "with an ambition for expansion and an accompanying tendency for expenses to outrun receipts," and a conservative party opposed to growth and change.[44] There does not seem to have been any such permanent alignment, even though temporary coalitions were effected in 1901. Along a different line entirely was the charge by an engineer in 1885 that engineering societies, "from the humblest society of workmen, up to the A.S.M.E. itself," had a tendency for two or three men to gain control and effectively discourage discussion.[45]

---

[42] "Annual Meeting of the American Society of Mechanical Engineers," *Amer. Mach.*, December 12, 1901, p. 1380.

[43] "Reports of Special Committee on Rules and Methods," ASME, *Trans.*, XXIV (1902–3), 889–920.

[44] Hutton, *History*, pp. 210–11.

[45] James F. Hobart, "Engineer's Societies—Small Things—Babbitting Boxes—That Hog-Nose Drill," *Amer. Mach.*, August 1, 1885, p. 3.

Another engineer wrote *Engineering Magazine* in 1893 that coteries which formed in mechanical engineering societies tended to stifle activity and new ideas and to make criticism unwelcome.[46]

A study of the presidents of the ASME over the period from 1880 to 1915 reveals that the nominating committee in effect elected three marine engineers, four educators, six consulting engineers, and twenty-three manufacturers, works owners, and managers of shops. This indicates that in terms of the role played as mechanical engineers, the society was not plagued by cliques, but was ruled by one big one. Of course, this distribution is not incompatible with the actual membership of the society. One could, however, question the likelihood that such a distribution represented the occupation as a whole.[47]

If cliques existed, they were mostly temporary and lacked the kind of strength necessary to challenge the ruling oligarchy; they also probably existed within that very ruling group. Some disagreement existed, for example, over the practical usefulness that papers given before the society should have. Some felt that they should be scientific and represent seminal contributions to knowledge; others felt that the interchange of practical information was the highest function of the society and that the perfect paper was a short talk on a practical problem, followed by a long discussion by members of their own experiences with this problem. In fact, the matter was adjudicated by having both kinds of papers. This conflict was to some extent cast within the context of school versus shop and was a disagreement between professors and shop managers, but there was no clear-cut division of opinion along these lines.[48]

The question which cut across the lines of the school-and-shop controversy, and went beyond it, was one of theory versus practice. *American Machinist* reported in 1899 that those who

---

[46] Albert D. Pentz, "Machine Shop Practice," *Engineering Magazine* (New York), VI (1893–94), 254.

[47] List and discussion of ASME presidents in Hutton, *History*, pp. 78–79.

[48] For criticism of the type of papers given before the ASME, see "Report of New York Meeting, November, 1883," ASME, *Trans.*, V (1883–84), 18; "The American Society of Mechanical Engineers," *Mechanical Engineer*, May 21, 1881, p. 160; "A New Departure in the American Society of Mechanical Engineers," *Mechanical Engineer*, April 28, 1883, p. 160; Editorial, *Mechanics*, November 11, 1882, pp. 305–7.

attended the ASME meetings were subjected to loss of time and patience by being forced to listen (apparently out of formal session, unless some of the discussion of papers was not reported) to bragging contests between practical men who thought knowledge could be acquired only through greasy overalls and some technical graduates who felt that only their kind should be given responsibility in connection with any engineering work. The "unoffending and well-disposed persons attending in the line of duty or to gain instruction are obliged to witness egregious strutting and listen to ignorant and more or less cantankerous misrepresentations of each side by the other, besides having their time wasted."[49]

Fred Halsey made a slight stir in 1898 when he suggested that the younger men in the society were being "sat down upon" too often. A reservoir of talent was being wasted, and there was "considerable feeling" about the problem among the younger element in the society.[50] Again, the problem came from within the ruling group, and it was met at once by the establishment of junior meetings where the younger men could gain experience and confidence.

Beyond its primary function as an agency for the presentation and publication of papers and treatises in the practice and theory of mechanical engineering, the ASME engaged in remarkably little activity of a professional nature. The society did not take action in the first twenty years of its existence to make clear to the larger society just what it regarded as the systematic, scientific knowledge base for the profession; to give clear directives to the educational institutions regarding the specific training expected of the college-graduated mechanical engineer; to standardize methods of entry into the profession; to regulate the use of titles; to take a decisive and clear

49 "The 'Practical Man' and the 'Engineer,'" *Amer. Mach.*, September 29, 1898, p. 730. Evidence that considerable material was deleted from the content of meetings as reported in the *Transactions* can be found in ASME Council Minutes, November 5, 1880; June 28, 1881, moved and passed that the Publishing Committee "be instructed to print only such papers, and discussions, as shall, in its opinion, be valuable"; November 4, 1881; June 1, August 8, and October 26, 1883, for removal of discussions involving "personalities"; and January 20, 1891, discussion of letter from Oberlin Smith, objecting to editing of his presidential address!

50 "Proceedings of the New York Meeting," ASME, *Trans.*, XX (1898–99), 23.

position on the question of limitation of self-interest; or to develop a code of ethics. In short, for at least the first twenty years, and perhaps even the first thirty, the society did not function as a professional association, but primarily as a clearing house for practical and scientific information and an elite social club.[51] This unwillingness to come to grips with the basic issues of professional status and behavior is nowhere more evident than in the disinclination of the society to use its influence to assure adoption of industrial and engineering standards of many types.

It was the practice of the ASME and, as will be shown, was consistent with the ideology and practice of the shop elite themselves, not to take public stands on various issues and to avoid being used as a forum for new and untried schemes. A very conservative attitude was taken toward any statement that was to go out under the society's imprimatur and thus bear the sanction of the nation's mechanical engineers. This sort of caution is understandable and was probably necessary. Gus Henning made a public protest to the society concerning a member who claimed to represent the society and had given antimetric testimony before a congressional committee. His motion to make a specific rule against this unauthorized use of the society's name and prestige was passed quickly.[52] Excessive caution, however, led inevitably to inaction.

Most of the proposed endorsements were presented to the society by one or more members, then voted on viva voce. It was almost impossible to get the society to pass such resolutions since the discussion preceding a vote always led to the conclusion that it was better not to stick out the society's collective neck. These resolutions, on such questions as the metric system, government test commissions, reform of the patent office, improvement of the status of naval engineers, were, according to F. R. Hutton, not the sort of thing that carried

51 Note ASME Council Minutes, December 7, 1900, relating to refusal to consider boiler standards; February 15 and May 4, 1883, concerning rejection of a proposal that the ASME establish uniform standards for collegiate degrees in technological courses; and the decision November 2, 1893, that any setting of standards for types of degrees given was "outside the immediate function of the Society."

52 "Proceedings of the Boston Meeting," ASME, *Trans.*, XXIII (1901–2), 433–35.

weight with congressmen anyway. More effective were personal letters sent to congressmen who were friends of eminent engineers. Personal, private action was preferred to institutional, public action.[53]

Stationary engineers' organizations were quite active in trying to secure laws which required operators of steam boilers or steam engines to pass examinations to determine their competence. The ASME, however, dissociated itself from any attempts to formulate proposals for general license laws for engineers, particularly for professional mechanical engineers. In 1910 the society took the rare step of writing a petition to the New York State Legislature opposing a bill "demanding the requirement of a license before any person could practice surveying or by implication follow other lines of engineering."[54]

The *American Machinist* printed a sample of a stationary steam engineer's examination in 1885, suggesting that few mechanical engineers or any other kind could answer all the questions.[55] Older, experienced engineers, one correspondent felt, would have little chance against the "graduate of some technical school" with all the data fresh in his head.[56] Two ASME members were on the committee of five appointed by the Franklin Institute in 1883 to study the whole question of licensing of engineers. The committee (including Washington Jones and Coleman Sellers) gave as reasons for opposing license laws that they were in restraint of liberty, the most important qualification (sobriety) could not be tested, cheating was easy, and most important the "passage of such an ordinance would create a privileged class of men, with the power to fix their own wages, regardless of the services rendered; and as proprietors would be compelled by law to employ only those having a license, it would assist an odious feature of trade's unionism."[57] This feeling was typical of the shop culture reaction to license laws. The licensing of mechanical engineers was never brought

---

53 Hutton, *History*, pp. 58–60. See ASME Council Minutes, May 26, 1885, for decision-making process regarding standards.

54 Hutton, *History*, p. 122.

55 "Examination for Stationary Engineers," *Amer. Mach.*, April 25, 1885, p. 8.

56 W. H. Wiggin, Letter to the Editor, *Amer. Mach.*, August 15, 1885, p. 5.

57 "Examinations of Stationary Engineers in Philadelphia," *Amer. Mach.*, February 9, 1884, p. 7.

up or discussed at an ASME meeting from 1880 to 1910. That stationary engineers (from whom the mechanical engineers wished to be distinguished) pushed so hard for license laws and allowed the issue to dominate their organizational efforts may partly explain the avoidance of the issue by the ASME. Mechanical engineers definitely objected to the idea of regulating who should practice engineering, but beyond this most license laws involved the other side of the coin, the regulation of the employers and entrepreneurs who hired them. Neither the arbitrary determination of who was to work as an engineer, nor who one could hire to do an engineer's work would have appealed to the engineer-entrepreneurs of the shop. The opposition which mechanical engineers exhibited toward license laws reveals, moreover, that the membership and leadership of the ASME were unable to agree on what might constitute a universal knowledge base to be expected of all mechanical engineers, upon which tests could be administered to identify and eliminate quackery and total incompetence.

After the discovery at the turn of the century that the society was heavily in debt and that its organizational procedures were out of step with the latest methods of business and industry, there was feeling, particularly among some of the younger members, that some change was necessary. This feeling was intensified by the impending move in 1906 to a new Engineering Societies Building. The first step was financial and clerical reorganization, spearheaded by ASME president and scientific management expert Frederick Winslow Taylor, and planned and executed, in large part, by Taylor's young assistant Morris L. Cooke.[58] More significant than the changes effected was the care which the reorganizers took to avoid offending such old shop culture deans as Henry W. Towne (responsible for the original system) and ASME secretary Frederick Remsen Hutton, who had served the society since 1883. Hutton was induced to retire, and a search was conducted for a new secretary who could implement the new system for running the society, and in effect bring it into the twentieth century. The new secretary,

58 See Morris L. Cooke's letters to Charles W. Baker, H. H. Suplee, Fred J. Miller, F. R. Hutton, and F. W. Taylor, June 1, 1906, in Taylor Collection.

Calvin W. Rice, and the leaders in the reorganization worked hard to placate Hutton's feelings that he was being by-passed and that his methods were being discarded.[59]

Typical of the other changes that were made in the operation of the ASME was the establishment of sections based on technical specialties. Until the reorganization of 1904–8 the ASME resisted all attempts at dividing the meetings into sections based on the particular interests of different members. By the early 1900's, when over five hundred members were regularly attending the New York meetings, and when such specialties as internal combustion began to challenge steam as a major concern of mechanical engineers, the need for sections was finally admitted. The council was given life-and-death power over their existence, they were required to be self-supporting, and they could not endorse standards or devices in the society's name, but after 1904 they were recognized.[60] In spite of this, it was not until 1908 that, for the first time, the ASME split up into sections and went to different rooms to hear papers on such topics as gas power, machine shop practice, steam and power plants, and experimental data.[61]

In the first decade of the twentieth century changes occurred in the ASME which amounted to a revolution when compared to the stability and continuity of the first twenty years. Yet, this new group was coming into power with the same basic framework and with an allegiance to shop culture, so it was not an accession to influence of the technical school graduate over the shop man and engineer-entrepreneur. It was merely a group of younger men wresting the reins of the society from an older generation. Young, scientifically trained and oriented Fred Taylor, new ASME Secretary Calvin W. Rice, and Fred Halsey, among the innovators in society practice, were men with experi-

59 Morris L. Cooke to F. W. Taylor, August 20, 1906; F. W. Taylor to Fred J. Miller, June 7, 1906; letters written relative to the appointment of the new secretary indicate two new categories of major contenders for power—engineering educators and scientific management experts. Worcester R. Warner to F. W. Taylor, June 5, 1906, and Walter C. Kerr to F. W. Taylor, June 6, 1906; all in Taylor Collection.

60 "Proceedings of the New York Meeting," ASME, *Trans.*, XXVII (1905–6), 39–40. Demand for sections included those of geographic nature, such as those requested but denied to St. Louis and Schenectady. ASME Council Minutes, April 16, 1903.

61 "The Spring Meeting at Detroit," ASME, *Trans.*, XXX (1908), 107–22; "Annual Meeting," ASME, *Trans.*, XXX, 517–36.

ence in and reverence for shop culture. They made the society more responsive to the desire for growth and diversification without letting the control of the society's affairs get out of the hands of those whom they felt were most capable.[62]

With tight control of the society by the self-perpetuating council, it was possible vastly to increase the membership of the organization and to include the thousands of technical school graduates without diluting the power of the shop elite. From 1905 to 1923, under the aegis of the energetic Calvin Rice, ASME membership skyrocketed from 2,500 to nearly 18,000, and included more than half of the mechanical engineers in the country.[63] The society was able to become a mass membership organization—with attendant influence and financial support—without having to reduce substantially the power of the shop culture elite. This leadership did bow, by the end of the first decade of the twentieth century, to demands that the society take a more active role in professional activities and American society in general. The society began to endorse standards and to operate a professional register for job placement. By the next decade, however, new groups, which cut across engineering specialties, were formed and began to take over the elements of aggressive professionalization that the ASME was still unwilling to engage in.[64]

With minor exceptions, during the first twenty years of its existence the ASME did not function as a professional association. It did not do those things which are expected of the representative bodies of occupations in the process of becoming professions. The ASME did add to the prestige of mechanical engineering simply by its very existence and by the character

62 Frederick W. Taylor wrote to Calvin W. Rice, ASME secretary-elect, that he would "try to have my friends on the Nominating Committee appoint men who will be especially friendly to you, as it is of course highly desirable that you should have not only the backing of the Council but also its most cordial and friendly co-operation." Taylor to Rice, June 7, 1906, Taylor Collection. That there was fear in the ASME of being overrun by the great numbers of young college graduates is indicated by the proposal in 1909 that the age for associate member be raised from 26 to 30 years. ASME Council Minutes, May 28, 1909.

63 Calvin W. Rice, "Fifty Years of the A.S.M.E.," *Mechanical Engineering*, LII (April, 1930), 268. Rice included an enormous increase of membership among his goals as secretary. Fred J. Miller to F. W. Taylor, June 14, 1906, in Taylor Collection.

64 For a discussion of the engineering profession from 1900 to 1940, see Edwin T. Layton, "The American Engineering Profession and the Idea of Social Responsibility" (unpublished Ph.D. dissertation, University of California at Los Angeles, 1956).

and reputation of the men who led it. The implication is that the men who founded the ASME and were members and leaders in its early years were men secure in social status in American society and that they founded the organization to lend the occupation of mechanical engineer the status they already possessed as individuals. The ASME became a kind of gentlemen's club based in New York City, whose membership shared in common the fact that they were mechanical engineers. This conception of a gentlemen's club was lent support by such things as the absence of aggressive professional control either through licensing or otherwise; the emphasis placed upon the society "House" in New York, a building with living accommodations which got increasingly elegant throughout this period; the annoyed attitude toward the resistance to raising dues expressed by some members; the restrictions on membership; and the tight method of controlling the political organization. Indeed, the simple joy of associating with other men of one's own class who recognized and appreciated achievements in *engineering* (almost in spite of social prestige) may have been a major impetus in founding the ASME. Since it performed so well as an organization of gentlemen engineers, the members of the ASME would have been surprised at the suggestion that it was not performing vital professional functions. They would not have felt those additional functions necessary for the increase of their own status.[65]

Local engineering societies, even more social in nature and purpose than the ASME, including all types of engineers from civil to electrical, and usually growing out of old civil engineering societies, were different in many respects from the national body which represented only mechanical engineers. Most of them started earlier than the ASME, and many were formed by the late 1860's. It is particularly interesting to note that some of the most influential were formed first in the Midwest and that to some extent the movement started in the West and moved East. The Engineer's Club of St. Louis began in 1869,

[65] On the fraternal, gentlemen's club atmosphere of the formal and informal gatherings of ASME members, see John Fritz, *The Autobiography of John Fritz* (New York: John Wiley & Sons, 1912), pp. 227–65.

Zerab Colburn

Alexander Lyman Holley

William Sellers

Coleman Sellers

as did the Western Society of Engineers, with headquarters in Chicago. The Civil Engineers Club of Cleveland, however, was not formed until 1880, and the Boston Society of Engineers in 1874. Partly, this can be attributed to the fact that civil engineers were often the prime movers behind these organizations and the civil engineers were located where the railroads were being built. Partly, however, this must be attributed to a dissatisfaction with the way the eastern establishment was running the major societies.

Often begun as local engineers' clubs, these organizations, as in the case of the Western Society of Engineers, sometimes branched out to take in all types of engineers and attempted to become influential regional societies, rivaling the national societies in prestige and breadth of activity. The Western Society, after thirteen years as a local affair, enlarged its scope in 1880 and helped found the Association of Engineering Societies, a central coordinating body to edit and report the transactions and proceedings of the other regional societies.[66] Although the founders had plans for the AES to become a kind of master engineering society for all types of engineers from the whole nation, its functions were limited, by finances and the growth of the four major national societies, to the publication of proceedings and, after 1884, to the publication of an index to engineering literature in the technical periodicals. These two organizations, the WSE and the AES, both located in Chicago (the largest enclave of mechanical engineers outside the states of the eastern establishment) could have provided a serious threat to the national status of the newly founded, New York-based ASME.

This may explain the inclination of the ASME over the first twenty years to discourage local societies, primarily by refusing to affiliate with them. It was not until 1907, after the reorganization of the ASME, that any attempt was made to develop a workable affiliate policy.[67] When in 1880 a young draftsman wrote to the *American Machinist* suggesting a national society

66 "Western Society of Engineers, Transactions; Address of E. S. Chesbrough, Retiring President," Association of Engineering Societies, *Journal* (hereafter cited as AES, *Jour.*), I (1881–82), 196–204.
67 Hutton, *History*, pp. 290–91.

for younger, aspiring engineers, the editors advised him to try the local societies, but added that "the social and personal advantages arising from connection with the national organization could not, of course, be shared by the membership of local societies."[68]

By the 1880's most of the major local or regional societies gave equal opportunity for membership and leadership to the elite of all branches of engineering. The officers and executives of the Michigan Engineering Society in 1885 included mechanical, civil, and mining engineers.[69] This greater breadth of technical background, as well as the presence of civil engineers, who often held posts as municipal engineers, led to a much greater and much earlier interest in civic improvement in such areas as pure water supply, street paving and lighting, and sewers. This was particularly true of societies where civil engineers tended to dominate, such as the Engineer's Club of St. Louis, and was less true of societies where mechanical, mining, and metallurgical engineers tended to dominate, such as the Engineers' Society of Western Pennsylvania.[70]

Even when they purported to be truly regional, as did the Western Society of Engineers and the Engineering Association of the South, local societies tended, like the national societies, to have most of their membership in one city or small area. The Engineering Association of the South reported in 1890 that its membership was distributed as follows: "Tennessee 43 per cent, Kentucky 23 per cent, Alabama 10 per cent, Georgia 5 per cent," with the rest scattered among the other southern states. The headquarters of the society was in Nashville, Tennessee.[71]

---

68 June 12, 1880, p. 8. For the studiedly nonlocal attitude held by the ASME, see the decision of the council not to involve itself in local affairs when requested by the New York City Superintendent of Buildings to help in drawing up a new building code, October 27, 1896; the council similarly dragged its feet when asked by representatives of the Association of Engineering Societies to participate in joint publication efforts and pondered "the policy it is expedient for the National Society to pursue toward these local organizations." The council did, however, suggest in 1891 that an ASME meeting be held in San Francisco "to put beyond cavil the national character of the organization." February 17 and March 19, 1885, and June 16, 1891, ASME Council Minutes.

69 "Next Meeting of the Michigan Engineering Society," *Amer. Mach.*, January 23, 1886, p. 7.

70 Engineers' Society of Western Pennsylvania, *Pittsburgh: Commemorating the Fiftieth Anniversary of the Engineers' Society of Western Pennsylvania* (Pittsburgh: The Society, 1930).

71 John MacLeod, "President's Annual Address," Engineering Association of the South, *Selections from Papers Presented during the Fiscal Year Nov. 21, 1889 to Nov. 13, 1890* (Nashville, Tenn.: University Press, 1891), p. 37.

Local engineering societies, with the exception of the regional societies based in Chicago, did not indulge in substantially more professional activity than did the ASME. For the most part, they too functioned as gentlemen's clubs where engineers of the same background could meet and share experiences. They were more active in municipal affairs because of the local character of the membership and because of the large proportion of civil engineers.

Mechanical engineering periodicals frequently grew up with technical societies and played a complex and often inconsistent and ambiguous role in the internal development of the profession of mechanical engineering. Significantly, they did not exist before the beginning of professionalizing activity. Most of the mechanics' magazines of the period from 1825 to 1850 were gone by mid-century, and by 1860, when the *Engineer*, a short-lived Philadelphia journal, was founded, it could name only the *American Railroad Journal* (which had by that time become primarily a financial paper and had little engineering information), *American Railway Times* and *American Railway Review*, two publications similar to the *Journal*, and the *Scientific American* (then an illustrated patent journal with some success as a popular mass readership magazine) as the engineering journals then existing. The *Franklin Institute Journal*, in a class by itself, was, according to *Engineer*, respectable, "but its respectability is too much for it," implying that it was too scientific and not practical enough to furnish applied knowledge to and share experience with the shop milieu. Unfortunately *Engineer* itself soon folded and the mechanical engineers of the country were left without a voice or a conscience in print.[72]

Although a number of short-lived periodicals were founded between 1860 and 1876, and some like *Technologist*, coming out of Philadelphia from 1870 to 1876, genuinely addressed themselves to the problems of the mechanical engineer, it was in the period following the Centennial Exhibition of 1876 that a host of long-lived, widely circulated, and financially successful technical journals were founded. One does not necessarily have to look for conditions internal to mechanical engineering to explain this phenomenon. The quarter-century from 1875 to

72 "Engineering Literature," p. 49.

1900 saw the introduction and widespread use of high-speed rotary printing presses run by steam power, the linotype machine, and cheap wood-pulp paper. It also saw the expansion of the mail system through increased use of the American railroad network. It was thus possible for even a poor journal to survive in 1885 and for the best to fail in 1835. By 1910 there were several hundred technical journals representing every conceivable editorial position and specialized subject matter.

First in quality and scope among the post-1876 journals was the *American Machinist*, founded in 1877 by Jackson Bailey in New York. Although machinists did read it, the journal was and remained for about twenty-five years the major technical periodical for the mechanical engineering profession. Its editor was one of the prime movers of the ASME in its formative years, and the connection between journal and society was so close as to draw some criticism. *Mechanical Engineer* remarked in 1881 that it and other journals were not given information about the time and place of ASME meetings in time to print it. The journal sarcastically implied that the reason was because they "did not run the society."[73] Bailey recognized the conflict of interest possible in his close connection with the society and he was instrumental in getting a ruling that no editors be appointed to society committees or to the council. This mollified the critics somewhat and after a few years the issue was a dead one and technical editors were appointed and elected to society offices.[74]

Even though the *American Machinist* represented and defended the best of shop culture, its position was not consistent. This was probably because technical journalists chose to act as gadflies and conscience to the growing profession of mechanical engineering. This is clearly indicated in the case of the metric system. The journal took a position completely opposed to the demands of shop culture. Controversial opinions brought letters, subscriptions, and financial success to the *American Machinist*, but after a fight had raged in its columns for a year

[73] "The American Society of Mechanical Engineers," *Mechanical Engineer*, October 22, 1881, p. 134.
[74] Hutton, *History*, p. 39.

or so, the editors generally made a mediating statement that neither black nor white was true; instead the truth looked to them a trifle gray. The journal provided a vehicle for the information exchange characteristic of shop culture and a forum for the discussion and attempted resolution of the professional problems that plagued the mechanical engineer.

The multiplication and proliferation of technical journals continued apace, however, and by the mid-nineties, specialization was beginning to raise questions about the feasibility of the general shop journal. *American Machinist* in 1896 announced a new specialization toward the problems of machine construction.[75] By 1908, the ASME began publishing its own journal, *Mechanical Engineering*, and the *American Machinist* had lost its place as the forum and conscience for the profession. Nevertheless, its editor Fred J. Miller went on to become ASME president in 1920.

Another journal, *American Engineer*, published in Chicago from 1880 to 1890, took school culture as its common denominator and engaged in editorial battles with the shop culture oriented *American Machinist*. It gave much greater emphasis to the status of the engineer and the problem of becoming more professionalized, extolled the virtues of the technical graduates at the expense of the shop elite, and attacked journals like *American Machinist* for their "semi-engineering character" and their tendencies to rely on "old and time-worn shop kinks and wrinkles."[76] *American Engineer* accused the New York journal of catering to men who gloried in titles but had not earned them, and it libeled the nontechnically trained shop elite by saying in 1885 that "it is the great drawback of men engaged in a profession, who have not had the benefit of systematic training, that they are so apt to take part in, and advocate fallacies which the young graduate, however inexperienced, will never attempt."[77] Such a strong editorial stand against shop culture apparently did not do much for circulation, and late in 1885 a shift took place in the emphasis of the journal, the edi-

[75] January 2, 1896, p. 1.

[76] *Amer. Engr.* (Chicago), August 10, 1883, p. 62.

[77] "The Wrongful Use of Engineering Degrees," *Amer. Engr.* (Chicago), August 13, 1885, p. 61.

torial position was played down and finally abandoned, and the practical workshop section was expanded. This capitulation did not help, and the journal failed in 1890. If it had survived into the nineties, its message might have found a ready audience among the thousands of young technical graduates of that and the following decades.

Conditions were changing, however, and the broadly based journal with strong editorial opinion was being subtly replaced by the highly specialized technical journal with emphasis on information and not on conscience and general problems of professionalization. *American Machinist* recognized the trend by 1896. In that same year a Cleveland journal, *Scientific Machinist,* was forced by economic necessity to narrow its subject matter. The journal changed from a general orientation toward all branches of steam engineering and machine shop work to the specific interests of the "engineer-machinist" in the steam power plant. This had been done "recognizing the fact that this is an age of specialization."[78] By the early twentieth century the technical journal had a firm and secure financial future, but increasing specialization had robbed it of the chance to play a role in the internal development of the mechanical engineering profession. But from at least 1875 to 1905 it was a major formulator of the group issues represented by the aggregate private troubles of the individual practitioners.

[78] November 15, 1896, p. 4.

To understand the lack of professional activity on the part of both national and local engineering societies, one must look to less formal, but no less potent, influences on internal professional development or its inhibition. Role is one of these internal factors. This study is concerned with the patterns of employment and behavior among men playing occupational roles. Because shop culture so long dominated the employment market for mechanical engineers, the role they played tended to be entrepreneurial at some stage of their careers, and often quite early. Owing to low capital needs for entry into such industries as machine-tool building, the proprietor stage did not have to be the end result of a long career; in fact, many men started partnerships at an early age. This agrees with the finding of other researchers that entrepreneurs tended to have their first success before thirty years of age, as opposed to bureaucrats, who usually had theirs after age thirty.[1]

Other mechanical engineers had quite different histories. Many held positions as draftsmen, designers, managers, foremen, superintendents, and chief engineers in machine and engine shops. A sizable number were in the Naval Engineer Corps, a substantially bureaucratic context, and carried their feeling for bureaucracy with them when they left the Navy and went into the ranks of the engineering educators. Education, particularly after 1870, provided a possible career for young mechanical engineers and offered the kind of security of role that the naval service was losing. If, however, all mechanical engineers were either entrepreneurs or would-be entrepreneurs or bureaucrats, where was the independent consultant who so frequently forms the keystone of the ideological arch of professionalism? Was there a role in American industry for the independent professional adviser on mechanical problems? With some exceptions, it appears there was not.

---

[1] See charts and discussion in Reinhard Bendix, *Work and Authority in Industry* (New York: John Riley & Co., 1956), pp. 229–35.

Certainly there were the giants of the profession like John C. Hoadley and Charles W. Copeland who operated as consulting engineers, but even they often were simultaneously sales or commission agents for various kinds of mechanical devices. Such an obvious conflict of interests was not conducive to the formation of unbiased, independent opinions. An editorial in the *American Machinist* in 1896 attacked this ambiguous role and the use by "so-called consulting engineers" of a title which indicated professional, objective advice when in fact they were salesmen and not really engineers.[2] Although this article was in effect attacking a later development and certainly did not mean to include the true consultants, the fact remains that the actual percentage of mechanical engineers acting in the role of independent consultants never rose above three per cent of the practitioners.[3]

To the young mechanical engineer who wished to be neither entrepreneur nor bureaucrat, further discouragement was offered by the attitude apparently prevalent among manufacturers that one paid for outside engineering advice only if it proved to be useful. It was very difficult to secure payment for negative opinions on the feasibility of producing and marketing an invention.[4] The status of the mechanical engineer as a professional was not secure enough to enable many to play an independent role.

Consulting practice was possible, however, for those who were already playing other roles in the profession. Robert H. Thurston and many other engineering educators, for example, were called upon by industry to do consulting work of all types.[5] Those entrepreneurial engineers who formed or worked

2 "Commissions and Rake-offs—Consulting Engineers or Salesmen: Which?" *Amer. Mach.*, January 16, 1896, p. 94.

3 Table of occupations of Sibley College alumni to 1904, American Institute of Electrical Engineers, *Transactions* (hereafter cited as AIEE, *Trans.*), XXVI (1907), 1439. Alex Holley had a lofty vision of his role as an independent consultant. He wrote one client that "I am so much interested in making your plant a notable improvement, that I shall be compensated by the *results!*" Holley to H. W. Rathbone, June 8, 1878, in Holley Letterbooks.

4 "Consulting Engineer's Practice," *Amer. Engr.* (Chicago), March 27, 1885, p. 146. Alex Holley spent his own time and money investigating steel processes in Europe but had a hard time getting manufacturers to consistently subscribe to his series of reports and thus to support the work. Holley to Bethlehem Iron Co., December 26, 1877; Holley to James Park, Jr., April 24, 1877; both in Holley Letterbooks.

5 Thurston, "Our Progress in Mechanical Engineering," ASME, *Trans.*, II (1881), 429.

for a *firm* of consulting engineers could also play a successful role. Good examples of this are Westinghouse, Church and Kerr (a late nineteenth-century firm of consulting and contracting engineers which undertook large-scale erection of machinery)[6] and the firm of Lockwood-Greene in New England. David Whitman, known throughout the textile country as the "mill doctor," got a reputation for his ability to solve problems of mill construction and equipment. Amos Lockwood and Stephen Greene continued the business Whitman had started, but Lockwood had been a successful mill operator and entrepreneur before taking up consulting work. This type of consulting was very much like an elaborate version of the old millwright's role.[7]

More common than that of the outside mill engineer was the role of permanent engineer to a factory or mill. It became common for a young technical graduate to be hired by a factory as the "mechanical engineer." His role was to be a sort of general trouble-shooter and on-the-spot expert, but sometimes the duties were quite trivial and often included running or at least keeping an eye on the plant's steam engine. H. L. Gantt, scientific management expert, told the ASME in 1910 that the mechanical engineer was becoming a fixture in kinds of plants where he had never been seen before. Gantt particularly mentioned the textile industry but stressed that while the mechanical engineer had been, up to that time, primarily in power and metal working, he was now needed and wanted in all types of industrial activity.[8] One thing was significantly different about this type of employment: it lacked the contact with other mechanical engineers or even with considerable numbers of skilled mechanics that the metal-working shops offered. On the other hand, it was the kind of job which technical graduates with no shop culture or eastern establishment connections could obtain.

There was also some question as to whether this kind of work fit the capabilities of the mechanical engineer as problem-

6 A. W. Smith, *Kerr*, pp. 42–63.

7 Samuel B. Lincoln, *Lockwood Greene—The History of an Engineering Business, 1832–1958* (Brattleboro, Vt.: Stephen Greene Press, 1960).

8 "The Mechanical Engineer and the Textile Industry," ASME, *Trans.*, XXXII (1910), 499–506.

solver and innovator, or whether such jobs relegated him to a role of high-grade handyman. Alexander L. Holley was the example most often used to illustrate the idealized role of mechanical engineer extraordinary. As James C. Bayles spoke of him shortly after his death in 1882, "I regard Holley's mechanical talent as eminently practical and characteristically American. He ever sought 'convenient means' to facilitate, to simplify, to save labor, to economize where economy was profitable."[9] A more glorious vision was described by the journal *Engineer* in 1898: "The man who made the first horseshoe was an engineer, but those who made them afterwards were mechanics."[10] Nevertheless, neither of these conceptions of the mechanical engineer's role was necessarily violated by a role as a general mill engineer. In fact, the ideal engineer for the metal-working shops was envisioned as a sort of jack-of-all-trades and master of all.

C. W. Bennett thus described the work of the chief engineer of a steel plant in 1900:

The chief engineer will be aided by specialists, whose detail work is assimilated into a synoptic form to be passed upon or put into practical use [in the larger plant]. In the lesser plant the chief may be expected to deal intimately with any technicality that operative conditions may provide. He not only presides over the draughting board, handles the transit and level in laying out switch tracks, conduits, etc., but lays out and supervises the installation of electric plant complete with engines, generators, switch-boards, motor and light circuits.[11]

The role expected of the mechanical engineer of a plant or factory might include that of employee-relations counselor in charge of preventing or heading off strikes and other labor disturbances. Engineers and manufacturers frequently advised young mechanical engineers to gain the confidence of possible ringleaders of the workmen. By his knowledge of the men and their moods, the plant engineer was in a position to accomplish miracles in industrial relations, according to some observers.[12]

9 "A Tribute to Alexander Lyman Holley, April 19, 1882," ASME, *Trans.*, III (1882), 44.

10 "The Engineer and the Mechanic," *Engineer* (N.Y.), August 15, 1898, pp. 190–91.

11 "The Commercial Engineer," *Wisc. Engr.*, IV (January, 1900), 31–44.

12 For example, see Frank L. McVey, "Some Suggestions Concerning Strikes," Minn. Engr., *Year Book*, V (1897), 6; Robert H. Thurston, "The Mechanical Engineer—His Work and His Policy," ASME, *Trans.*, IV (1882–83), 96.

In contrast with these notions of the role of the mechanical engineer as universal problem solver and adjudicator of the industrial system was the idea of the mechanical engineer as specialist. Representatives of shop culture and school alike warned the young student engineer that he must specialize in order to survive. George W. Melville, Chief Engineer of the Navy, wrote in 1902 that during the preceding twenty-five years, "it became evident to the engineering executive officials that the field of industrial science, like that of law, medicine and diplomacy would no longer be grasped in its entirety by one mind." Engineers had seen that the rewards were to the specialist, and both the engineering profession and the world had gained from this change.[13]

Certainly there were changes occurring in the field of mechanical engineering which made such statements seem plausible, in spite of the fact that many engineers were becoming jack-of-all-trades mill engineers. Much of the new role of the mechanical engineer resulted from the fact that his basic subject matter was changing. Previously the engineer who knew a fair amount about steam engines and machine tools could play almost any role in industry. By the late nineteenth and early twentieth centuries, such subjects as gas engines, automobiles, airplanes, specialized conveying and hoisting machinery, elevators, and other internal machinery for tall buildings were coming within the field of the mechanical engineer, and most demanded the competence of specialists.[14]

In addition to this kind of specialization of new fields, the older field of steam was being broken up into such specialties as heating and ventilating engineering and power plant engineering.[15] Other engineering industries, technically outside the direct competence of mechanical engineers, began in the 1890's to hire young mechanical engineering technical graduates. The Chicago Telephone Company was doing this by 1905, under special training programs which included both electrical and

---

[13] "Naval Engineering Advance, Its Influence upon Sea Power," *Amer. Mach.*, November 6, 1902, pp. 1572–75.

[14] "The Changes in the Scope of Mechanical Engineering," *Amer. Mach.*, April 27, 1899, p. 345.

[15] Carl C. Thomas in *California Journal of Technology*, reprinted under "The Field of the Mechanical Engineer," *Amer. Mach.*, March 17, 1904, pp. 351–52.

mechanical engineers. The director of the program referred to them both as "telephone engineers."[16] As we have seen, the large electrical manufacturing firms such as Westinghouse and General Electric also had special training programs for young technical graduates in mechanical engineering which would enable them to develop particular competence in one kind of work.

Increasing specialization went hand in hand with increasing bureaucratization of employment. The large telephone and electrical manufacturing industries fit this pattern, and so did the railroads in the late nineteenth and early twentieth centuries. President Eckley B. Coxe told the ASME in 1894 that in consequence of the consolidation of railroads there were now more assistant and fewer chief engineers. Many more engineers were now forced into subordinate positions, from which there was little chance of promotion, and the upward movement that was possible was much slower than before. This was coupled with a "very decided and growing tendency to place the work, which was formerly in [the] charge of men who had risen from the ranks and become foremen, in the hands of graduates of technical schools."[17] Roy V. Wright, writing in the *Minnesota Engineer* in 1909, admitted that for some time technical school graduates had been hired for the drafting room at low salaries, were generally dissatisfied with the subordinate role they played there, and did not stay long. Wright felt that men from the correspondence schools, manual training schools, and night schools were more satisfactory since they did not "have such an exalted idea of themselves."[18] Thus, unless a mechanical engineer had executive ability, he had to be willing to take a definitely subordinate role. Nevertheless, many young graduates who had no choice continued to enter such roles and to take such positions. What they could expect was outlined by an engineering professor in 1896. H. Wade Hibbard wrote:

In a small office with six or eight draftsmen the chief draftsman or mechanical engineer usually deals directly with each subordinate; with fifteen or

[16] Angus S. Hibbard, "The Relations of the Engineer to the Telephone Business," *Wisc. Engr.*, IX (April, 1905), 176.

[17] "Technical Education," ASME, *Trans.*, XV (1893–94), 661.

[18] "The Mechanical Engineer and the Railroads," *Minn. Engr.*, XVII (March, 1909), 137–40.

twenty draftsmen three or four are leading men, the chief directing often through them, all the designs finally coming up to him for approval. To avoid errors a second draftsman checks up the work of the original designer and is held equally responsible with him for mistakes to an extent variable with the nature of the drawing. The chief draftsman then signs it as correct, and forwards it in case it is a drawing sufficiently extensive or standard as to require the approval of the superintendent.[19]

The professors of engineering schools were understandably reluctant to send graduates to such a fate, but they were not averse to suggesting a possible role for their students in stationary engineering. One writer in the alumni and students' magazine of the University of Wisconsin engineering school warned of a technical school graduate explosion in 1901, suggesting that sheer oversupply of engineers would force many into stationary power plant work. And why not, the writer cheerfully added. Such work was ideal, offering a situation in which theory and practice "operate together to produce a result." Salaries were going up and would ultimately be the highest for any kind of engineering. Even this sanguine exponent of stationary engineering admitted that many college-trained men who went into such work were disappointed. Wiping, oiling, and stoking simply did not add up to the "average engineering college man's aspirations."[20]

Along with the operation of steam plants and drafting, roles open to the technical graduate included teaching below the college level. Engineering instructors were needed for manual training high schools, part-time and night technical institutes, but such jobs, like drafting and stationary engineering, offered little in compensation or comfort to the professional ego. Nevertheless, school administrators wanted—and got—young technical graduates for these jobs. Most of these jobs in stationary engineering, drafting departments organized along bureaucratic lines, and high school teaching were not jobs that members of the shop culture elite would have taken or cared for. The young technical graduates may not have cared for them, but they were often forced to take them.[21]

---

19 "Railway Mechanical Engineering," Minn. Engr., *Year Book*, IV (1896), 110.

20 "Technical Graduates in Power Plants," *Wisc. Engr.*, V (May, 1901), 215.

21 W. A. Richards, "Manual Training in High Schools," *Wisc. Engr.*, VI (February, 1902), 172.

John Chipman Hoadley

Frederick Winslow Taylor

John Edson Sweet

Robert Henry Thurston

Frequently, subordinate positions within bureaucratic organizations and also within entrepreneurial shops where mechanical engineers owned no part of the business led to situations where the engineer had responsibility for the success of an operation without corresponding control over the operative conditions necessary for that success. Such establishments were often those in which the top executive was a man unskilled in the mechanic arts. It was particularly true of situations where experts in management science (who were also mechanical engineers) were called in to make a plant more efficient. Often the top executive would interfere with the work of the engineer and override his plans, forcing him to quit.[22]

Many young technical graduates were hired into the sales departments of firms manufacturing technical equipment. Their specialized knowledge was of value in selling complex and expensive engineering equipment. With only a few exceptions, engineers and technical journals agreed that this movement was good, and never was there a suggestion that a job as a salesman threatened the status of the professional or particularly complicated the role which the mechanical engineer played. On the contrary, most commentators felt that the true role of the engineer was in the business end of manufacturing since the actual technological break-throughs were nearly complete.[23]

One source of ideological and actual conflict within the industrial system was between technical graduates and intermediate managers, foremen, and top executives who had made their way to the top by personal skill rather than by formally attained qualifications or credentials. The uneducated did hire the educated, but often, like Thomas A. Edison, bragged about how they were hiring (at low salaries) men with high pretensions and only marginally useful abilities. Some, like J. S. Walker, president of the Engineering Association of the South, believed that emphasis on engineering as a learned profession was getting out of hand and was actually making it more difficult for engineers to play an effective role in industry. He told

---

[22] "Responsibility without Control," *Amer. Mach.*, March 10, 1892, p. 8; *ibid.*, April 7, 1892, p. 8; Guido Sacerdote, "Is the Right Man in the Right Place?" *Amer. Mach.*, November 16, 1911, pp. 927–28.

[23] C. L. Redfield, "What is a Mechanical Engineer?" *Amer. Mach.*, July 20, 1893, p. 6.

the association in 1898: "The continual claims for engineering as a learned profession may interest those who discuss it; but to put it mildly, it is bad policy, seeing that it tends to make those who employ us think that we feel ourselves to be wiser than they." Walker pointed out that the net result was that "the practical man hires the educated engineer as one of his aids, and employs him in a subordinate capacity, regarding his pay in the light of a very unpleasant item in the expense account."[24]

Others disagreed and wrote to technical journals that "the 'cut-and-try' old-style practical man is passing, and no one can blame the manufacturer who prefers the one who can give him exact figures to the one who 'will get it out somehow.' "[25] The ideological evidence is here so contradictory that it is impossible to derive fact from it. As is so frequently true, the ideological statements assume the existence of rigid monoliths, called variously "industry," "business," or "the trade," which either were or were not accepting and welcoming technical graduates and providing them with a role. In fact, a changing and expanding industrial structure was providing new kinds of roles for the new kinds of inexperienced mechanical engineers coming from the technical schools.

Some indication of the types of role change that occurred can be found by comparing a list of 650 graduates of all technical schools made up in 1896 with a list of 1,260 graduates of Cornell's Sibley College in 1904 (see Tables 3 and 4). Though obviously not comparable, the figures are suggestive and bear repeating here. Both lists indicate the numbers and percentages of graduates in each type of occupation. Beginning about 1887 there is a striking rise in new categories, new roles, which were not held by the engineers of previous classes. These include chemists, electricians, examiners in the patent office, and a tremendous growth from 21 per cent to 56 per cent between 1871 and 1895 of the "others" category, indicating that large numbers of mechanical engineering graduates were taking roles

24 "Annual Address of the President," p. 75.

25 Johnse, Letter to the Editor, "The Really Practical Man," *Amer. Mach.*, December 1, 1904, p. 1604.

TABLE 3: Occupations of Engineering Graduates by Class in 1896

Occupation	1871	1872	1873	1874	1875	1876	1877	1878	1879	1880	1881	1882	1883	1884	1885	1886	1887	1888	1889	1890	1891	1892	1893	1894	1895
Partner in Business	37	50	16	23	62	35	23	29	35	7	29	32	28	37	16	31	23	21	9	6	9	3		2	2
Manager in Business			16	7	6	10	5	17	10	14	6	11	6	13	15	7	6	3	6	6	6	3	6	3	2
Superintendent	6		8	8	6	6	5	5		7	23	11		13	16	10	7					3	2	2	
Mechanical Engineers	6								5								7			12	3				
Assistant C.E. and M.E.	12	6	12	8		23	24	6	10	21	12	3	11	8	4	3	10	9		12	6	9	6	9	2
Assistant Superintendent																									
Draftsmen			6			6	5	12	5	7	6	11	17	8	4	7		12	6	12	9	3	2	3	2
Professors and Instructors	6	12	6													3			16	12		21	12	13	28
Chief Engineer	6	6	6	15		12	5	6	10	7		6	11	4	8	17	23	18	12		10	3	7	13	3
Master Mechanic											6	7		4			3	3					2		
Chemists																		3	18	9	6	6	8	3	7
Electricians												4		4				9		6	3	3	5		
Examiner, U.S. Patent Office																	10								
Assistant Manager																			6	3	3	12	6		
Others	21	26	30	39	26	8	28	25	25	37	18	18	27	9	37	22	18	22	27	21	42	32	42	52	56
Total*	16	16	16	13	16	17	21	17	20	14	17	28	18	24	25	29	30	33	33	33	31	32	49	47	47

* Figures are actual numbers, not percentages.

TABLE 4: Occupations of Sibley College Alumni in 1904

Occupation	Number	Per Cent
Mechanical Engineer	298	23.20
Electrical Engineer	170	13.23
Designer or Draftsman	140	10.90
President, Vice-President, Secretary, Treasurer, or Member of Firm, Manufacturing	127	9.88
Teacher	114	8.88
Sales Engineer	107	8.33
Consulting Engineer	43	3.35
Manager or Superintendent, Operating	41	3.19
Nonengineering Occupations	39	3.03
Manager or Superintendent, Manufacturing	32	2.48
Foreman	31	2.41
Manager or Superintendent, Constructing	25	1.94
Insurance Engineer	19	1.48
Attorney	18	1.40
Army or Navy Officer	17	1.32
Editor or Publisher	15	1.17
Assayer, Geologist, or Mining Engineer	15	1.17
President, Vice-President, Secretary, Treasurer, or Partner, Constructing	13	1.01
Patent Examiner	8	0.62
President, Vice-President, Secretary, Treasurer, or Partner, Operating	7	0.54
Civil Engineer	4	0.31
Irrigating Engineer	1	0.08
City Engineer	1	0.08
Total	1285	100.00

in business and industry and elsewhere for which they were not specifically trained and which would not likely gain them entrance into professional associations or engineers' clubs. Easily explainable also is the rather neat rise from zero to 28 per cent of those holding jobs as draftsmen. This was clearly the way in which at least one-third began their careers by 1895. Whether it was for the earlier period, these figures cannot indicate. Such roles as mechanical engineer, assistant mechanical engineer, professor and instructor, and superintendent are fairly constant over the years and show no significant variation.[26]

The number of all graduates up to 1896 who were members or partners in firms was about 20 per cent of the total, whereas

[26] The preceding and following statistics can be found in a table of technical graduates, 1871–75, by percentage in each occupation, in Mark Talcott, "Admission Requirements to Engineering Schools," *Amer. Mach.*, June 4, 1896, p. 567, and in a table of present occupations of alumni of Sibley College, by number and percentage, in "Proposed Code of Ethics," AIEE, *Trans.*, XXVI (1907), 1439.

in 1904 this figure had dropped to a little over 10 per cent. Such fields as drafting and teaching remained about the same, each about 10 per cent. The general classification of mechanical engineer, however, rose from about 10 per cent to 23.2 per cent of the total. New categories appeared in 1904, such as insurance engineer with 1.48 per cent of the total and sales engineer with 8.3 per cent.

Although the category of city engineer was listed in 1904, only one individual out of the total of 1,285 graduates held such a position. The work of the mechanical engineer had not (at least by 1904) led him into the area of municipal problems as had that of the civil engineer and architect. This fact will be important to remember when considering the rejection by most mechanical engineers of a "large" role in society.

These statistics suggest that college-trained engineers from top eastern schools were becoming less inclined to enter situations in which they themselves would be entrepreneurs and were more likely to enter relatively large, bureaucratic corporations where their titles were specified as mechanical engineer and where they bore no immediate interest in the profit and loss of the organization. This was made possible by the intense development of technical education and by changes in industry which created new opportunities and roles for mechanical engineers.

The technical graduate without shop culture connections, without distinctions of birth, had no built-in status. His role, frequently more subordinate than that of the older generation of mechanical engineers had been, offered him little social status and an often unimportant occupational status in the industrial system. Because his new role frequently lacked manifest prestige and professional character, he and the engineering educators who had spawned him sought, in ways never used by the old shop-culture elite, to give professional status to the appellation of mechanical engineer, especially as represented by the technical school graduate bearing that title by academic fiat.

By the late nineteenth century even the small machine and engine shops began to change. Incorporation, standardization of many types of machine tools, and large demands for those tools turned many a small machine shop into a large manufacturing plant, and much of the personal contact and cooperation typical of the earliest shops disappeared. What had been bold experimentation in machine tools in the era from 1850 to 1875 often became stodgy conservatism by the 1890's. A good example was that of a very gifted shop engineer, Charles Norton, who was forced to leave Brown and Sharpe in order to experiment with production-grinding machines (destined to replace lathes, milling machines, and planers for many types of industrial operations) since the conservative management of the firm did not believe production grinding worthwhile.[27]

In shops grown conservative the mechanical engineer could play a less innovative role than he had enjoyed earlier. The bureaucratically oriented, college-trained engineer might find a place in some of these manufacturing companies which had once been a vital part of shop culture. It would be a mistake, however, to believe that shop culture had died by 1910. New, smaller shops appeared to carry on the innovation and to provide a role for the engineer who preferred designing to manufacturing.

[27] Robert S. Woodbury, *History of the Grinding Machine* (Cambridge, Mass.: Technology Press, 1959), pp. 97–105.

W. A. Truesdell told the readers of the *Wisconsin Engineer* in 1899 that the only way for engineers to achieve high social and occupational status was to become independent practitioners. He was willing to admit, however, that "the majority of engineers will be obliged to occupy salaried positions, and though there is a marked improvement in progress over what has been heretofore, they can never obtain that independence or security from abrupt changes which those will have who are in positions of their own." In short, the status of the salaried mechanical engineer was not secure.[1] The writer was wrong in assuming that previous generations of mechanical engineers had been consultants, but the question naturally arises—What of the status concern of the old shop-culture elite?

These men were and had always been quite secure in their status, deriving it as they did from a variety of sources: upper-class birth, entrepreneurial situation, frequently a classical education, knowing the right people, and last, but certainly not least, personal genius or intelligence. Oberlin Smith told the ASME in his presidential address in 1890 that the engineer must be both a scholar and a gentleman. It was important that the engineer have a classical education, know literature and philosophy, so that he could associate in business and socially with the people who counted.[2] Some of the most avid defenders of shop culture were both scholars and gentlemen. W. F. Durfee, who caustically attacked technical education, was a distinguished student of antiquarian Americana and wrote learned treatises on the history of ancient technology.[3] Alexander Lyman Holley had similar inclinations toward classical scholarship and also was a stout advocate of shop culture against the claims of the school. He was able, moreover, to make fast friends with upper-class European engineers and sci-

---

1 "A Talk to Young Engineers," *Wisc. Engr.*, III (January, 1899), 30–33.

2 "President's Annual Address: The Engineer as a Scholar and a Gentleman," pp. 45–47.

3 "How the Ancients Moved Heavy Masses," *Engineering Magazine* (N.Y.), VI (1893–94), 611–25.

entists on his many trips abroad.[4] Oberlin Smith had suggested, perhaps unwittingly, the class he was talking about by stating that the engineer should be able to discuss and converse with scholars and scientists "at your club." To assume that every mechanical engineer belonged to a "club" at which he would come into contact with "scholars and scientists" was quite an assumption indeed, but not an impossible one for an ASME president in 1890.[5] By 1905 such an assumption could not be held as reasonable.

The social functions of the ASME were not extensive but those they had were elegant affairs, always at the best hotels and restaurants, with receiving lines for the engineers and their wives and sumptuous many-course dinners. When on several occasions the ASME was invited to send an official party to Europe to visit international exhibitions, entire steamships were chartered, as in 1889 when the *City of Richmond* carried the cream of the mechanical engineering elite to Europe for the Paris Exhibition. These men were secure socially and showed it.[6]

It is no surprise then that they were not bothered by their status as businessmen and industrialists, nor did they feel that it threatened their status as engineers. Possibly some even regarded status as an entrepreneur as higher than that of an engineer. *American Machinist* noted top positions possible to the young man in the following order: "mechanics, draftsmen, engineers or proprietors."[7] Leading engineers frequently spoke of the "business" of mechanical engineering, and while not every manufacturer was necessarily an engineer, being an entrepreneur certainly did not prevent one from being an engineer. Compatible with this conception was the fact that the engineer-entrepreneur as a type was extolled and encouraged by engineers and manufacturers of all kinds.[8]

Many top mechanical engineers began and continued their

4 "The Engineers of the Old World Greet the Engineers of the New," *Amer. Mach.*, June 27, 1889, p. 7.
5 "President's Annual Address: The Engineer as a Scholar and a Gentleman," p. 45.
6 Hutton, *History*, pp. 226–27.
7 "College or Machine Shop," *Amer. Mach.*, January 16, 1890, p. 8.
8 "The Status of Engineers," *Mechanical Engineer*, August 9, 1884, p. 186.

careers in this dual role, and its attractiveness may well have been what led them into the field in the first place. Any attempt to deprecate the entrepreneurial spirit and to distinguish the engineer as a social and occupational entity separate from and superior to the entrepreneur was not likely to find favor among such men. If successful, such a move would discredit the leadership of the mechanical engineering profession.

Even though they felt no immediate need for increased status, elite mechanical engineers did seek certain measures which would have had the effect of enhancing that status. One of the pet schemes of mechanical engineers from Oliver Evans to the leaders of the ASME was the creation of a government post of cabinet rank to deal with mechanical engineering. Such a department would have control over all mechanical and engineering work done for the government, replacing the various informal ways in which such work was contracted. If this idea contained seeds of lofty professionalism, its implementation was very questionable.[9]

More within the realm of feasible action was the effort to raise the status of the mechanical engineer as an expert in court testimony. Mechanical engineers were being called to testify as experts in boiler explosions and other areas of their competence at least by the 1870's. What so often happened, however, was that each side hired its own mechanical engineers who naturally gave differing opinions, giving both judge and jury about as high an opinion of the professional ethics and competence of mechanical engineers as they now frequently hold regarding psychiatrists under similar circumstances. In one such case in 1880, one side spent between $4,000 and $5,000 hiring experts, running the cost of the trial up to $1,500 per day.[10]

Many engineers were troubled by the reluctance of courts to accept evidence read from an engineering book. The technical journals often ran editorials on the problem, sometimes even suggesting that the best answer was for engineers to refuse to testify at all and thus avoid the chance that they might com-

9 "A Department of Mechanics and Engineering," *Amer. Mach.*, August 18, 1892, p. 8; "A Department of Mechanics and Engineering," *Amer. Mach.*, May 11, 1893, p. 8.
10 "Value of Expert Evidence," *Amer. Mach.*, May 1, 1880, p. 3.

promise themselves and thereby lower the profession in the eyes of the public.[11] Particularly in patent cases the question of objectivity was a crucial one.[12] Gradually the courts recognized mechanical engineers as experts, and after a test case in New York in 1898, even the use of published engineering tables and other works was allowed in evidence, providing "such statistics and tabulations are generally relied upon by experts in the particular field of the mechanic arts with which such statistics and tabulations are concerned."[13] Such rulings greatly strengthened the status of the mechanical engineer as a professional.

Engineers objected, however, when professional status was used as a justification for public or private bodies asking for free engineering advice. Such was the case in 1885 when a Citizens Committee of Grand Rapids, Michigan, seeking advice on a sewerage problem, requested that an engineer give them free advice as a matter of "professional interest." It was pointed out in *American Engineer*, which discussed the incident, that engineering stood as high as medicine and law as a profession and that these professionals would be compensated for such advice.[14] Engineering manufacturers also found themselves occasionally in the position of giving free engineering advice to people and firms, and they complained about it.[15]

There was some concern among the technical journals at least that mechanical engineers were passed over or otherwise ill treated when it came to appointments to public commissions. An editorial in *Engineer* complained that no engineer was appointed to the commission to investigate the causes of the *Maine* disaster in 1898.[16] An anonymous correspondent of *Mechanical Engineer* noted with disdain that one Thomas Shaw, a manufacturer of engineers' goods, was defeated for the position of Chief Engineer in the Philadelphia Water Department "because he would not stoop to the tricks of petty politicians."[17]

11 See *Amer. Engr.* (Chicago), February 27, 1885, p. 98.
12 George Escol Sellers, "Early Engineering Reminiscences," *Amer. Mach.*, January 23, 1890, p. 4.
13 "Use of Authorities by Expert Witnesses," *Amer. Mach.*, June 23, 1898, p. 473.
14 October 15, 1888, p. 152.
15 "The Passing of Free Engineering Advice," *Amer. Mach.*, August 29, 1907, p. 315.
16 "An Engineer's Duty," *Engineer* (N.Y.), March 1, 1898, p. 58.
17 "Politics in Professions," *Mechanical Engineer*, April 14, 1883, p. 91.

A *cause célèbre* of this sort came in 1893 in connection with the Chicago Exposition. The affair was set off when the Director of Works of the fair declared at a dinner in New York that "in the last decade or two one could go out on the streets of a great city and collect a force of engineers and draftsmen very much as formerly he hired mechanics."[18] H. F. J. Porter answered the Director's remarks in the *American Machinist* in May of 1893. The reason for the subordination of engineers at the exposition was, he declared, because engineers were not organized like members of the profession of architecture. Architects of ability and reputation were *sought*, but engineers had to *apply* for positions and were not so well compensated.[19]

The editors of the *American Machinist* carried the subject further in July, reiterating the idea that the architects had fared better in their relations with the exposition because they had a "strong professional union" similar to a trade-union. By contrast, the mechanical engineer had no "fixed minimum fee for services or advice; no rules or code of ethics, and in fact, little or nothing to protect members from the effects of unrestricted competition for work."[20] Porter enlarged his original statements at an ASME meeting at the exposition in August of 1893. The trouble was that the engineering schools were not turning out a uniform product. Porter felt it was necessary "both for the good of the profession and for public safety, not only that a high grade but also that the same grade of men should be turned out from the technical schools." The engineering profession, he told the ASME, lacked status and was in a disorganized condition, thus making its members helpless before the better-organized segments of society.[21] Status, according to such thinkers, was dependent not so much upon the accomplishments as upon the assertiveness of the individual engineer and the profession as a whole.

---

18 "Engineers and Mechanics," *Amer. Mach.*, April 20, 1893, p. 8.

19 H. F. J. Porter, "The Status of Engineers," *Amer. Mach.*, May 4, 1893, p. 6. H. F. J. Porter, son of Civil War General Fitz-John Porter, was educated in private schools in New Hampshire and at Lehigh University. See Fred V. Larkin in *DAB s.v.* "Porter, Holbrook Fitz-John."

20 Editorial, "Status of Engineers," *Amer. Mach.*, July 20, 1893, p. 8.

21 Comment on Thurston, "Technical Education in the United States," ASME, *Trans.*, XIV (1892–93), 1005–6.

As an editorial in the *American Engineer* expressed it, " 'self-assertion' is the watchword needed to bring about a reform in the standing of the engineer; for no one cause has operated so powerfully to lower the profession in popular estimation as the lack of this quality."[22] The same journal on another occasion laid to lack of self-assertion the blame for the prevention of engineers from "assuming the social position, and the emoluments which accrue to the other professions and trades."[23] George W. Melville, Engineer-in-Chief of the Navy, wrote in 1902 that the engineer had been, over the preceding twenty-five years, reluctant in demanding recognition of his achievements.[24]

A correlative of the lack of self-assertion was the lack of general public recognition of the status of the engineer. Not that all mechanical engineers believed this to be so, but enough did to be able to make a sizable commotion in the technical press and elsewhere about it.[25] One engineer wrote the *American Machinist* in 1897 that "people—the masses—have no idea what engineering means. They don't know it requires just as fine an intellect to be a civil or mechanical engineer, as it does to be a lawyer or doctor." He combined this idea with the concept that every successful engineer must have a technical education.[26] The answer was partly in educating the public to understand the work of the engineer and to an appreciation of him as a professional. While it would be impossible to say that such concern was never entertained by the shop culture elite, it is possible to say that engineering educators did express it, and, as will be shown later, another largely bureaucratic group, the naval engineers, worried about this problem very much.

One theme which both engineering educators and naval engineers stressed was the lack of appreciation of the status of the engineer and his works in literature. Professor J. A. L. Waddell told the graduating class of Rose Polytechnic Institute in 1902 that both English and American novelists "sneered at the engi-

22 "The Standing of the Engineer," *Amer. Engr.* (Chicago), December 1, 1882, p. 254.
23 "The Standing of the Engineer," *Amer. Engr.* (Chicago), June 17, 1882, pp. 282–83.
24 "Naval Engineering Advance, Its Influence upon Sea Power," pp. 1572–75.
25 George W. Dickie, "The World's Injustice to the Technical Man," *Amer. Mach.*, November 7, 1895, p. 891.
26 A. L. Bowen, "Technical Education," *Amer. Mach.*, November 25, 1897, p. 880.

neer, terming him a 'greasy mechanic' and placing him outside the pale of polite society."[27] ASME President and Chief of Naval Engineers Charles H. Loring told the assembled mechanical engineers in 1892 that

contemporaneous historians have but scantily drawn attention to the immense influence exerted upon modern history by the steam-engine. They follow in the same well-worn ruts, giving dubious descriptions of battles, names of monarchs, lists of decrees and laws, no end of political negotiations and intrigues, and the whole array of puppets who seem to push the car of time, while they are only flies upon its wheels. The real shaping cause of the march of modern events and of the great industrial progress of the times has but trivial recognition in the literature which pretends to account for what has happened, or to predict what may ensue.[28]

Some engineers disagreed, like J. S. Walker of the Engineering Association of the South, who felt the profession of engineering got all it deserved "in the way of praise and regard."[29] Speaking before the graduating class of Clarkson Institute of Technology, William McClellan said in 1909 that those engineers who were broad-gauge in their abilities and conception of the role of their profession "are not complaining about recognition by the community. It knows their names, and is proud of them."[30] Few members of the ASME leadership actually felt deeply enough about alleged lack of public recognition to say or do anything about it.

Similarly, they were unconcerned about the problem of whether engineering as a profession was compromised by the existence of engineering as a trade or business. Early in the growth of technical education the engineering educators were inclined to agree with them that this was no problem. By 1911, however, some defenders of school culture were beginning to become concerned over apparent status difficulties encountered by their graduates. J. A. L. Waddell wrote in 1911 that in the

[27] J. A. L. Waddell, "Address to the Members of the Graduating Class in the Engineering Department of the Rose Polytechnic Institute," *Addresses to Engineering Students*, ed. John A. L. Waddell and John L. Harrington (Kansas City: Waddell and Harrington, 1911), pp. 422–23.
[28] "President's Annual Address: The Steam Engineer in Modern Civilization," ASME, *Trans.*, XIV (1892–93), 255.
[29] "Annual Address of the President," p. 68.
[30] "The Engineer and the Community," *Addresses*, ed. Waddell and Harrington, p. 393.

preceding generations "it was claimed, even by some engineers, that engineering was not a profession but merely a trade . . . but today all that is changed," primarily because engineering was becoming the most important profession of all. This movement could be furthered by making certain that engineers were sufficiently paid, protecting the engineers' rights by law, establishing a code of ethics, and gaining the general respect of the public. He discussed in detail a plan afoot to create one powerful, national, professional organization with very high standards, specifically that a member be forty years of age, a degree-holder, and speak and write one foreign language.[31]

If engineering was a thing apart from common trades, it is but one step further to assume, as did one correspondent of the *American Machinist*, that "many occupations might be mentioned which it would not profit an engineer to engage in, even though he might thereby acquire several times as many dollars per year as his profession procures for him."[32] Such conceptions of engineering as a unique calling sometimes led to flights of fancy concerning the status of the engineer. Colonel H. G. Prout told the graduating class of Rensselaer in 1901 that "when we mass the contributions of the engineer to modern society, when we consider the physical, the intellectual and moral results of his work, we must conclude that his is the noblest of the professions."[33] Frederick H. Bass, assistant professor of engineering at the University of Minnesota, believed that the key to why engineering was a profession and not a trade was the concept of "service to mankind."[34] Francis C. Shenehon, dean of the College of Engineering at Minnesota, could bring himself to call the engineer "co-partner with the gods."[35]

Status for the engineer as professional obviously went beyond the boundaries of any one engineering specialty. Very infrequently did engineers or educators speak of the professional

31 "The Present Status of the Engineering Profession and How It May Be Improved," *Addresses*, ed. Waddell and Harrington, pp. 281–89.

32 G. E. Flanagan, "The Need of Practical and Technical Training," *Amer. Mach.*, August 21, 1902, p. 1200.

33 "The Engineer's the Noblest of the Professions," *Amer. Mach.*, August 22, 1901, pp. 945–46.

34 "The Status of the Engineer," *Minn. Engr.*, XIX (January, 1911), 71.

35 "Advice to Freshmen," *Addresses*, ed. Waddell and Harrington, p. 21.

qualities of mechanical as against civil, mining, and electrical engineering. Civil engineers did make such distinctions, as we shall see, but mechanical engineers did not. Nevertheless, concern for professional status was often linked with the idea that to remain a professional the engineer should attack and expose amateurs and charlatans wherever he found them. Robert H. Thurston, no doubt with some sarcasm, wrote in 1889 that America was the home of freedom and "exhibits one aspect of 'freedom' which, to the unsophisticated citizen of the old world is more extraordinary than perhaps any other among the thousand singular developments of liberty which here attract his attention; this is the freedom accorded to the amateur to enter into the serious work of the professional."[36]

American Machinist agreed with Thurston's indictment of the concept of every man as his own engineer, at least inasmuch as it applied to certain specific cases. The editors on several occasions cited cases where lawyers, bankers, and legislators had tried to supervise everything from the building of water works to the design of steam warships. "There is no department of engineering," they wrote, "which the layman hesitates to enter."[37] *American Engineer,* a Chicago-based periodical which was very much concerned about the status of the engineer as professional, deprecated employers who acted contrary to the advice of professional engineers and who even tried to hold engineers responsible for the invariably bad results. The journal advised engineers to go so far as to quit rather than work under such conditions. Only this kind of autonomy would bring about the desired result, that "the engineer should be accorded an authority in his department similar to that which is now generally accorded the physician in his special profession."[38]

One of the key issues in the winning of professional status by engineers who felt they did not have it was the use of titles. All professionally oriented mechanical engineers agreed that the use of the general title "engineer" needed a complete over-

---

[36] "Amateur Engineering," *Amer. Mach.*, December 26, 1889, p. 7.
[37] *Amer. Mach.*, March 27, 1890, p. 7.
[38] July 1, 1882, p. 6.

hauling. In the eyes of the public, the word could mean anything from a locomotive operator, to an engine oiler, to the President of the ASME. As Oberlin Smith told the ASME, "It is certainly a bad thing that every fellow who, perhaps, cannot read or write, but who can grease a little agricultural engine, should in this country be called an 'engineer.' "[39] An anonymous correspondent of the *American Machinist* noted that "if you say you are an 'engineer' nine out of every ten persons will think that you operate a steam engine."[40] Indeed, even the usually aloof ASME considered the question seriously, and several papers were given on the subject.[41] It was not, as the *American Machinist* noted humorously, after one such paper and discussion, a simple matter of snobbery. "There are those who seem to think that the 'Engineers,' (i.e. the Mechanical Engineers), at that meeting objected particularly to bearing the same title borne by 'engineers' (i.e. stationary engineers), or 'engineers' (i.e. locomotive engineers), or 'engineers' (i.e. engineers employed in paper mills), because these latter wear overalls and sometimes have soiled hands."[42] In spite of the democratic nature of shop culture ideology, even established members of the shop culture elite did not care to be mistaken for engine drivers, even if, in principle, they could not mind being mistaken for shop mechanics.

The obvious answer was either to force the engine drivers and other subengineer groups (from the viewpoint of professional mechanical engineers) to describe themselves in another fashion, such as "engine driver" (this was frequently proposed), or to change the appellation of professional mechanical engineers (this was also proposed). Both suggestions required that the public be made to understand and accept these changes, and both required that some official body demand the change. Both conditions were unlikely, and the proposal in the *Railroad Gazette* that "Engineer" be left to the locomotive operator and

---

39 Comment on Woodward, "Training of a Dynamic Engineer," ASME, *Trans.*, VII (1885–86), 770–71.

40 "The Draftsman and His Work," *Amer. Mach.*, December 18, 1902, p. 1832.

41 H. F. J. Porter, "Topical Discussions and Interchange of Data, No. 525–96," ASME, *Trans.*, XIV (1892–93), 487–527.

42 "The Title of Engineer," *Amer. Mach.*, March 9, 1893, p. 8.

the term "Ingeneer" be applied to the professionals died a-borning.[43]

Going beyond the general term "engineer" to the new prefix produced similar, if less traumatic, dilemmas. Although the term "mechanical engineer" was in common use in this country by the 1850's, both Robert H. Thurston and Alex Holley used their school-granted title of civil engineer for the first several years of the existence of the ASME. Many noncollege-trained engineers were at first a bit dismayed by the very pretentiousness of the ASME itself. One prominent member was quoted by the *American Machinist* as saying "that he had never yet, in all his business career, written the initials 'M.E.' after his name, nor called himself a mechanical engineer, his highest ambition having been to be a 'pump maker.' "[44]

There was some disagreement over whether the prefix "mechanical" was the proper one at all. At the time of the founding of the ASME, a small group, led by Professor William P. Trowbridge of Yale, tried to get the name changed to "dynamical engineer." This change was rejected by an overwhelming majority, but the controversy continued, with several major engineering schools graduating "dynamical engineers" until about the turn of the century.[45] *American Engineer* surmised that the fight in the ASME was stimulated by a feeling of the Trowbridge group that "the word 'mechanical' conveyed the idea of 'routine like,' 'without thought or knowledge,' 'like a machine.' "[46]

The serious conflict over the use of titles centered on the question of whether the title "Mechanical Engineer" should be reserved for degree-holding technical graduates, or whether it should be reserved for those who had proved their capabilities as engineers, degree-holders or not, or whether both groups should share the title. William Kent maintained that only college graduates should use the title.[47] Academically oriented periodi-

---

[43] Quoted in "Ingeneer and Engineer," *Amer. Mach.*, October 1, 1903, p. 1382.

[44] September 18, 1880, p. 8.

[45] Hutton, *History*, pp. 10–11; *Mechanical News, An Illustrated Journal of Manufacturing, Engineering, Milling, and Mining* (hereafter cited as *Mechanical News*), July 15, 1886, p. 136.

[46] "Civil and Mechanical Engineering," *Amer. Engr.* (Chicago), I (October, 1880), 153.

[47] Editorial, *Amer. Engr.* (Chicago), September 17, 1888, pp. 111–12.

Main machine room of the Jarecki Manufacturing Co., Erie, Pa., 1881 (*Scientific American*).

cals like *American Engineer* took a similar position in editorial statements that "M.E. should *only* be used by those who have pursued a certain course and passed certain examinations in a Polytechnic Institute."[48] Use of the title by nondegree-holders was fraud, no matter how brilliant or successful they might be. The journal went so far as to suggest that it was primarily a matter of courtesy and that able men who were not technical graduates did not assume the title.[49] This last statement was simply not true. In the 1880's, when this controversy was at its bitterest, most members of the ASME signed themselves "M.E." and most were not technical school graduates.

Opposition to this idea was quickly forthcoming from the traditional defenders of shop culture and its prerogatives, such as the journal *American Machinist*. Kent's distinction was absurd, according to that journal, since it made "M.E." mean only those with little or no experience.[50] *Mechanical News* joined the fray with the comment that such arbitrary distinctions would make it possible for "college graduates who may or may not know anything practical about engineering [to] enter upon the undivided possession of the 'title.' "[51] *Scientific Machinist* stated the shop culture case in clear terms in an 1896 editorial:

Engineering is a science that may be learned in the shop and from the private study of books. Many of the most competent engineers are men whose education was procured entirely outside of the engineering schools. No one entrusts engineering work of importance to a man until he has proved himself capable. A title is absolutely no guarantee of capability, and if men not graduated from any scientific school should be compelled to omit the letters M.E. or C.E. from their names, the title business would suffer.[52]

Engineering educators differed in their reaction to hard-core resistance by noncollege-trained mechanical engineers to the attempted theft of their newly won title. Some, like Robert Thurston, changed from giving bachelor of science in engineer-

48 Editorial, *Amer. Engr.* (Chicago), I (September, 1880), 134.
49 Editorial, *Amer. Engr.* (Chicago), December 15, 1882, p. 271.
50 Discussed in *Amer. Engr.* (Chicago), September 17, 1888, pp. 111–12.
51 July 15, 1886, p. 136.
52 "Would Protection Protect Engineers," *Scientific Machinist*, July 15, 1896, p. 4.

ing degrees and began giving the actual degree of "Mechanical Engineer."[53] Others, pressured by shop culture resistance, or merely realizing that all their graduates did not go into engineering, gave the B.S.M.E. or B.M.E. degree, frequently offering the title "M.E." to the graduate who went out and proved himself, then presented himself for knighthood.[54] Technical graduates often found that their pretentious "M.E.'s" did them little good without experience to back it up. One graduate wrote the *American Machinist* in 1903 after four years at a well-known eastern college that he was lucky to get a job doing drafting at a low salary. He soon ceased writing "M.E." after his name and now believed that no school had the right to grant this title. His own alma mater *now* awarded the Bachelor of Science in Mechanical Engineering.[55]

The latter degree did become the norm, and Thurston was bucking the trend in 1885 when he instituted the change in Sibley College from "B.M.E." to "M.E." However, such a degree fitted better in to his grand plan for the M.E. degree to be followed by graduate degrees, such as masters and doctors of engineering, which he hoped would ultimately raise the status of American mechanical engineers to the level of their continental European counterparts. Concern for the use and regulation of titles was exhibited much more by the defender of school culture than by the advocate of shop culture. This phenomenon was symptomatic of a general and serious concern for raising the status of the mechanical engineer which the elite often did not understand or appreciate, and which the educator found necessary for the ideological support of his new system.

The members and defenders of the shop culture complex did not worry about the status of the engineer because most of the individuals involved already had social and economic status of high order. If anything, they had lent prestige to mechanical engineering by their practice of it and had made it a "gentlemen's" profession. They resented college-trained engineers of

---

53 Thurston to J. G. Schurman, February 18, 1896, in Ex. Com. Papers; Durand, *Thurston*, p. 108.

54 Storm Bull, "Higher Degrees," *Wisc. Engr.*, VII (May, 1903), 214–15.

55 Grad, "Technical Education as a Graduate Sees It," *Amer. Mach.*, April 30, 1903, pp. 623–24.

both mediocre social and intellectual background. By contrast, the educators and their products exhibited increasing interest in raising the status of mechanical engineering as a profession. They did this because they saw it as one way to enable the college-trained engineer to receive the deference already accorded to the successful shop engineer with good social background. The educators believed that ultimately the fact of a man's being an engineering graduate should confer upon him the status which the shop culture elite enjoyed by virtue of factors external to its engineering activities.

CHAPTER 9

IN SEARCH OF A RATIONAL
SHOP ENVIRONMENT

Although the American mechanical engineer showed little interest in professional status, his occupational role made him ever interested in the problem of systematizing and standardizing the materials with which he worked, the methods he used to measure them, and the tools he used to work them into articulated machines. He sought, out of necessity, to give these things a systematic and rational basis which they lacked. This theme of rationalization united the sometimes discordant followers of shop and school cultures, united technical graduates and old shop masters; in short it was an internal development which provides a key to the increasingly professional orientation of the American mechanical engineer.

The first president of the ASME, Robert H. Thurston, said in his inaugural address that "we are to endeavor to hasten the approach of that great day when we shall have acquired a complete and symmetrical system of mechanical and scientific philosophy." The mechanical engineer must, according to Thurston, use simple, comprehensive, and effective methods to do his work with the "highest efficiency and economy."[1] Such concern could lead to a general call for national efficiency and rational use of the nation's resources, as when the *American Machinist* in 1882 stated that the "great mechanical and scientific problem of the future is the utilization of waste forces."[2] One obvious contribution the mechanical engineer could make in this realm was to rationalize the objects and processes with which he came into contact each day.

One distinct problem which bothered many mechanical engineers was the fact that steam boilers, despite precautions and increasingly stronger construction, continued to explode with some frequency and with considerable loss of life. Francis B. Stevens, superintendent of the Hoboken shops of the Camden

1 ASME, *Trans.*, I (1880), 11.
2 *Amer. Mach.*, June 17, 1882, p. 8.

and Amboy Railroad, succeeded in 1871 in persuading Congress to appropriate $100,000 to support the work of a scientific commission to investigate in a rational, experimental way the exact causes of boiler explosions. The next year Robert H. Thurston was brought in as a consultant, but he quickly noted various internal organizational difficulties which made it difficult to accomplish anything. By 1876 Thurston had resigned, the money was expended, and the results were meager. Thurston's chief complaint seems to have been that the commission was not systematic and thorough in its approach and failed utterly to produce publishable experimental data.[3]

Thurston was appointed Secretary of the Iron and Steel Board created under an act of Congress in 1875. He had full charge of this commission, was given an appropriation of $100,000, which was intended for the investigation of the "physical properties of iron and steel as structural materials." Promised help by the Naval Ordnance Bureau, much of which never materialized, the board suffered from lack of sufficient funds, and when no further appropriations were forthcoming, "died a natural death from lack of support." It did, however, publish two large volumes of experimental data. Working with Thurston on this project were A. L. Holley and William Kent.[4]

During the first twenty years of its existence, the ASME was the forum for a running discussion of these short-lived and generally unproductive commissions, and it was one center of a movement to persuade the government to make new appropriations to revive or re-create them. In this, however, the ASME was actually following the lead of the ASCE, the AIME, and the Iron and Steel Institute, which prior to 1880 had tried jointly and singly to persuade Congress to re-establish the testing commissions. There was almost universal agreement among engineers of all types that the government had a duty to support such experimental activities.[5]

One of the most comprehensive statements of the case for esting commissions was made by Oberlin Smith before the

[3] Durand, *Thurston*, pp. 77–78.

[4] *Ibid.*, pp. 79–80.

[5] Robert H. Thurston, comment on Oberlin Smith, "Experimental Mechanics," ASME, *Trans.*, II (1881), 64. See letter from Alexander L. Holley to Abraham S. Hewitt, July 14, 1876, in Holley Letterbooks.

ASME in 1881. The situation, according to Smith, demanded order and system in experiments and the elimination of duplicated effort. Experiments, to be useful, had to be finished and recorded and made accessible to the "mechanical public in a properly indexed form." Only the steam engine was "rationalized," Smith told the ASME, and the problem of "ascertaining by tentative methods the fitness, strengths and qualities of different materials" had never been systematically attempted because the profit concern in industry dictated that experiments be carried only as far as necessary to get specific required information. One could hope, Smith said, eventually to educate the government to the importance of engineering science and to the sponsoring of such research. Equally important was the creation of a central council for the systematic direction of experiment and research, whose personnel would include mathematicians, physicists, engineers, and mechanics of the highest ability. Smith hoped to see the day "when *one* well-done calculation or experiment shall replace a *thousand* half done, and system shall replace chaos."[6]

Smith's plea brought from Robert Thurston the whole story of his efforts with government commissions and with the railroad testing laboratory at Stevens, which had become, he claimed, more commercial and less scientific since other duties had prevented him from directing it personally. A. L. Holley spoke out against the narrowness of experiments conducted by the Naval Ordnance Department and told the ASME that they should try to persuade the Department to do experiments that would have wider value.[7]

The following year, in 1882, Thomas Egleston of the Columbia University School of Mines made a major address to the ASME on the topic of "The Appointment of a United States Commission of Tests of Metals, and Constructive Materials." Egleston spoke of opposition from "certain quarters that are well known," presumably meaning the Ordnance Department,

---

6 "Experimental Mechanics," pp. 55–59. Compare Smith's concept of a central research council for technology to a similar concept worked out by the American scientific community. See A. Hunter Dupree, *Science in the Federal Government* (Cambridge, Mass.: Harvard University Press, 1957), pp. 115–19.

7 Thurston's and Holley's comments on Oberlin Smith's "Experimental Mechanics," ASME, *Trans.*, II (1881), 65–67.

which he specifically attacked on several later occasions.[8] Although there is no evidence to indicate the cause of the difficulty between the Ordnance Department and the engineers, it can be surmised that one issue could have been the invasion by the engineers of the area of testing which the Ordnance Department regarded as its special competence. The ASME did appoint a committee to follow and encourage the progress through Congress of the perennial bills introduced on the subject of a testing commission. In addition, this committee sent circular letters to institutions and to individual manufacturers asking them to put pressure on Congress. The movement was unsuccessful, however, and the organization gradually lost interest in it as by the 1890's private industries and universities began to take up this type of basic research.[9]

More directly productive were the related efforts to formulate and enforce a uniform code of methods for testing so that the results of different experimenters would be comparable. Thomas Egleston was the prime mover in this movement, which succeeded in 1884 in establishing a committee of the ASME to pursue the matter systematically.[10]

By the November, 1891, meeting the committee had developed a detailed and systematic procedure for the testing of all types of materials used by mechanical engineers, and the society endorsed (by accepting the report) the work of the committee. By this time, however, the problem had become one of international importance and scope, and international commissions, to which the ASME sent delegates, were taking the matter in hand.[11]

The ASME and the technical journals were the vehicles for the expression of an interest in the rationalization of the nomenclature of the machine and the shop. One engineer, in a letter to the *American Machinist,* called for an end to "mechanical provincialisms" and the introduction of standard terms for machine parts; for example, the "six-square nut" should be called a "hexagon nut."[12] Oberlin Smith demanded such reform

8 ASME, *Trans.,* III (1882), 81.
9 ASME, *Trans.,* IV (1882–83), 31–32.
10 "Proceedings of the New York Meeting," ASME, *Trans.,* V (1884–85), 16–17.
11 "Proceedings of the New York Meeting," ASME, *Trans.,* XIII (1891–92), 23.
12 Olin Scott, "Mechanical Provincialisms," *Amer. Mach.,* April 9, 1881, pp. 6–7.

before the ASME in 1881. He attacked the usual practice of deriving one part name from another, which often led to such designations as a "lower-left-hand-cutting-blade-set-screw-lock nut." He proposed a comprehensive system of letters, numbers, and symbols which would rationalize the description of machine parts.[13] It was not, however, as a result of any action by the ASME or the technical journals that a systematic machine shop nomenclature developed; it was mainly the result of a generation of engineering textbooks and handbooks.

Concern for a "clean" and uncluttered shop vocabulary was related to the broader conception of utility as beauty in machine design. Designers of stationary and marine steam engines and locomotives before 1870 displayed considerable ambiguity regarding decoration and ornamentation, and frequently, polished metal, fancy painting, and Greek Revival and Gothic decoration appeared on such machines. Machine tools were from the first more straightforward, but, particularly in light woodworking machinery, ornamentation and decoration sometimes found expression. Beginning in the 1840's and 1850's with the development of basic designs for commercial sale, and culminating in the work of the Philadelphia toolmaker William Sellers in the 1860's, heavy machine tools such as gear cutters, planing machines, and large lathes took on an uncluttered, strictly functional appearance, and "machine gray" replaced fancy painting. This simplification of appearance went hand in hand with increasing complexity of operation and increasing precision of workmanship. Though Sellers himself did not publicly acknowledge a debt to an ideal of utility as beauty and vice versa, other engineers did articulate such sentiments in the years that followed.[14]

As early as 1860, the editor of *Engineer* saw Philadelphia-built locomotives possessing a special and distinguishing combination of form and function. *Engineer* maintained that "a truly beautiful locomotive—for extraneous ornament and skin deep

---

[13] "Nomenclature of Machine Details," ASME, *Trans.*, II (1881), 66–77; "Experimental Mechanics," p. 69.

[14] Bertold Buxbaum, "Der amerikanische Werkzeugmaschinenbau im 18. und 19 Jahrhundert," *Beiträge zur Geschichte der Technik und Industrie*, X (1920), 130; Coleman Sellers, Jr., and Alexander E. Outerbridge, "William Sellers," Franklin Institute, *Journal*, CLIX (May, 1905), 369–70.

decoration do not constitute beauty—must be the work of one who thoroughly understands its mechanism; and the machinist who can produce an elegant design will be constitutionally sensitive in the matter of workmanship, with all its mechanical refinements of fit and finish."[15] *Technologist* similarly noted that despite a common misconception that "elegance is allied to weakness," it was a fact that "almost all lines developed by a careful analysis of the forces required to resist strains are beautiful curves."[16] On another occasion, calling for "Simplicity in Design," the same journal praised machine designs which were consistent in the way "visible shaft, bolt, stud, and nut-endings [were] turned or milled." Not only was it more beautiful, but it avoided dirt-catching corners and made maintenance easier.[17]

By the time the ASME was founded in 1880, the idea had achieved some popularity among engineers. In a tribute to the late A. L. Holley in 1882, Coleman Sellers told the ASME that Holley had said to him, "Don't you know that you and I have had a good many talks about the abominable practice of putting different styles of architecture, Gothic and what not, into steam engines and other machines. We know perfectly well that what is the most fit is the most beautiful."[18] It was a concern of the shop culture elite because it fit into their conception of the role of the shop in the education and instruction of the young engineer in the mysteries of mechanical engineering. As F. F. Hemenway wrote the *American Machinist* in 1881:

I am of the opinion that a neatly designed, neatly finished and painted machine has a good moral effect on the man who has much to do with it. . . . The quality of beauty takes upon itself a new function—that of utility. . . . When I buy my machinery I want it to look well from an engineer's point of view, and I want it to look well from an artist's point of view. . . . When a man is to pass the best part of his time amongst his machines and machinery, he ought to arrange to get work and instruction therefrom—pleasure and profit should result.[19]

15 "Beauty in Locomotives," *Engineer* (Phila.), September 13, 1860, p. 33.
16 "The Relation of Elegance to Strength in the Designing of Machinery," *Technologist*, II (July, 1871), 169.
17 *Technologist*, II (August, 1871), 207.
18 ASME, *Trans.*, III (1882), 61.
19 June 18, 1881, p. 4.

As the *American Machinist* aptly put it, the proper punishment for critics of industrial society, like the Englishman John Ruskin, was to place them on an island with no machinery.[20] Many in the shop culture elite would have agreed with William T. Magruder of the Engineering Association of the South when he described the highest examples of the engineer's designing art as "startlingly true" and "in the category of the fine arts."[21] While they agreed that form and function went together, critics like C. L. Redfield expressed concern that there were no schools and no distinct profession of machine design.[22] Charles T. Porter, a prominent steam engineer, believed that one great lack of the profession was a true science of the strength of machine tools.[23] Again, it was the textbooks on machine design, such as those by Joshua Rose, that helped to bring about the general acceptance of the idea of the unity of form and function.

Oberlin Smith carried the idea of the uplifting character of the rational shop to its logical conclusion in an article he wrote in 1896. He stressed the effect the use of individual electric motors would have on the shop, eliminating "hangers, shafting, pulleys and belting." He described the future shop as having walls that were "smooth and clean, with a permanently light color, better than old-fashioned whitewash, to assist in reflecting the light. The windows will be clean and very numerous, assisted in their function by sky-lights." Smith waxed romantic and almost poetic when he came to describe the type of machinist who would inhabit this perfect shop:

In the ideal shop, above pictured, the work will be done by these clean-looking, active men, whose energies will be augmented by the electrical spirit prevalent in the air, and who will be encouraged to higher methods of thinking, working and living by their reasonable hours, good wages and

---

20 "Art's Hatred to Utility," *Amer. Mach.*, September 13, 1884, p. 3. In light of this comment, it is interesting to note Leo Marx's interpretation of Shakespeare's *The Tempest*. See his *The Machine in the Garden* (New York: Oxford University Press, 1964), pp. 34–72.

21 "Mechanics as a Fine Art," *Engr. Assoc. South, Trans.*, VI (August, 1894), 46. William Bement, the Philadelphia tool builder, was reported to have said "I think it useful to figure the strength of machine parts because the results are suggestive to the designer," quoted in A. W. Smith, *Sweet*, pp. 14–15.

22 "The Relation of Invention and Design to Mechanical Progress," *Engineering Magazine* (N.Y.), XII (November, 1896), 286–91.

23 "Strength in Machine Tools," ASME, *Trans.*, I (1880), 1–3.

care taken for their comfort and health—as well as by the uniform warmth, good ventilation, and bright pleasant light always surrounding them—with such environments, and not until we get many such, may we expectantly look for an advent of the coming *scientific machinist*.[24]

This kind of concern is, however, within the context of one aspect of the rationalization movement known as scientific management, which will be dealt with in a later chapter.

Interest was expressed by engineers, and action was taken by manufacturers on an individual firm level, to standardize such basic items as screw threads and pipe flanges. In the case of the screw thread, William Sellers was again the innovator. He rejected the British Whitworth standard and designed a standard thread of his own, which he began using exclusively for his machine tools. Due to his influence with the directors, the Pennsylvania Railroad followed suit by adopting the standard, forcing their suppliers to do so, and so on. Thus engineers who were also influential and strategically placed manufacturers were able to create standards and to force or persuade others to adhere to them.[25] Pratt and Whitney was another firm that took unilateral action to establish standards. This kind of action was applauded by most engineers. Nevertheless, there was considerable opposition to the government's or the ASME's adopting and trying to enforce standards arbitrarily.[26]

William Kent was the chairman of the committee on standard methods of conducting steam boiler tests which reported to the society in 1885, and he was also the individual who led the forces who did not want the society to formally adopt the committee report.[27] Kent's position prevailed and the report and future committee reports on standards were merely printed in the *Transactions* without comment. A heated debate raged over the question of whether the society should adopt and recommend standards over the next several decades, with the no-action group winning. The official reasons distilled from the debates on the subject included the following: it would not be

24 "The Scientific Machine Shop," *Scientific Machinist*, January 15, 1896, pp. 8–9.

25 William Sellers, "A System of Screw Threads and Nuts," Franklin Institute, *Journal*, LXXVII (May, 1864), 344; Roe, *Tool Builders*, pp. 248–49.

26 George M. Bond, "Standard Measurements," ASME, *Trans.*, II (1881), 81–92.

27 "Report of the Committee on Standard Pipe and Pipe Threads," ASME, *Trans.*, VIII (1886–87), 29–44.

proper for the minority of the society attending meetings to act for the rest viva voce—thus any such action should be by letter ballot to all members; few in the membership at large possessed the competence to judge a standard; adoption of a standard implied business partnership with those engaged in making and selling articles affected by the standard; a standard would be difficult to change once adopted; controversies over adoption within the society would give it the atmosphere of the market place; and the society might be held legally responsible for pecuniary damage done to those forced by its action to adopt the standard.[28]

Throughout its first thirty years, the society steadfastly held to this position of nonaction, despite concerted attempts by small groups within the society to adopt this or that particular standard. The conflict had diverse individuals on the advocacy side in each instance, and whatever it may have represented, it was not a simple conflict between school and shop.

One particular controversy, which occurred from 1888 to 1893, does seem to have been a school-shop conflict. In May, 1889, James W. See gave a paper before the ASME in which he attacked the lack of standards for such various items as carriage clips, washers, bricks, picture frames, needles, and files. Naming over a hundred such items, he called for the creation of a government bureau of standards which would record all standards (not merely of measurement, as was then being done, but of such things as screw threads, flange designs, and uniform sizes of industrial parts). The men of shop culture approved heartily. Oberlin Smith, Coleman Sellers, and Henry Towne all remarked in favor of pursuing the idea. However, Robert Thurston and other leaders of the school group opposed the idea with vigor.[29] This alignment makes sense when one considers that it was not proposed that the government bureau *establish* standards but merely that it *record* the standards set by such manufacturers as William Sellers and Pratt and Whitney so that they would be accessible. What would result would

28 Hutton, *History*, pp. 61–64.
29 James W. See, "Standards," ASME, *Trans.*, X (1888–89), 542–75; Discussion on article in "Proceedings of the New York Meeting," ASME, *Trans.*, XI (1889–90), 23–31.

be a free-for-all competition between different standards, all duly registered, and the best one would win out. Why such a market-place concept would not have appealed to the bureaucratically oriented Thurston is not hard to imagine. Oberlin Smith, as usual playing the role of mediator and visionary simultaneously, suggested amending See's original idea to include the setting of standards by the bureau, which would be "a competent 'syndicate' of specialists" selected with "the aid and advice of our leading technical colleges and engineering societies." This plan went far beyond what the rest of the shop culture elite would have liked, and See protested the perversion of his idea, though one may imagine that Smith's position brought him ever nearer to the camp of the centralizing, bureaucratic systematizers.[30]

Most engineers believed that it was desirable to have rationalized standards of measurement, nomenclature, fittings, screws, nuts, bolts, and everything else with which they came in daily contact. The difference between school and shop on this issue was over method. Engineering educators frequently spoke out for the centralized establishment of standards by governmental or engineering bodies—that is, for one set of standards to which all would conform. Naturally, they imagined themselves as the arbiters of these standards. Shop culture, on the other hand, believed the methods of the open market place would suffice. Individual shops and firms would develop standards and the best one would eventually drive the others from the field. The shop elite were opposed to the idea that any centralized body, including the ASME which they controlled, set and enforce a single standard for each item. The idea that someone should suffer pecuniary loss because of being forced to adhere to an arbitrarily set standard was objectionable to them; that someone suffered because of the inevitable and natural action of the market was unfortunate but acceptable. On one issue, however, shop culture went one step further and opposed rationalization itself as well as the methods employed to secure it.

More than any other such controversy the fight within the

---

[30] Oberlin Smith, comment on See, "Standards," ASME, *Trans.*, X (1888–89), 567–68.

ASME over the question of whether to adopt the metric system or to recommend that the government make it the United States standard reveals the complex and variant strains within American mechanical engineering. The campaign for the adoption of the metric system was by no means confined to the ASME. Indeed, it had begun in this country soon after the system's adoption in France during the French Revolution, and it has continued unabated until the present. The sources of support for and opposition to the system in the United States have not been systematically investigated, but some conclusions can be reached regarding the alignment of American mechanical engineers on this issue.

Coleman Sellers, active defender of shop culture, started the ASME on its antimetric course by making the question of its adoption for the machine shops of America the subject of the first formal paper given before the new society. Of all men, Sellers told his audience with questionable logic, mechanical engineers were the most appropriate group to address the question of metric adoption without bias, because they used weights and measures in their daily work and because they were concerned with the *"rationale"* behind any system they used. Sellers went on to make a very tightly reasoned technical case against the metric system based primarily upon the concept that the meter and the millimeter were less appropriate units to use in the shop than the foot and particularly the inch. He closed his remarks with an attack upon the "young engineer [who] enlisting himself in the ranks of the metrical reformers buys a cheap scientific notoriety." Sellers asked why engineers should try "something *new* but no *better* at so frightful a cost."[31]

The widespread approval of Sellers' position is significant in that it came primarily from the shop engineers and particularly from those who were simultaneously manufacturers. Henry R. Worthington, maker of pumps, rose to the occasion by saying, "For the first time in my life, I stand before a meeting that is

---

[31] "The Metric System: Is It Wise to Introduce It Into Our Machine Shops?" ASME, *Trans.*, I (1880), 2–19. See also J. E. Hilgard to Coleman Sellers, April 29, 1874, in Peale-Sellers Papers.

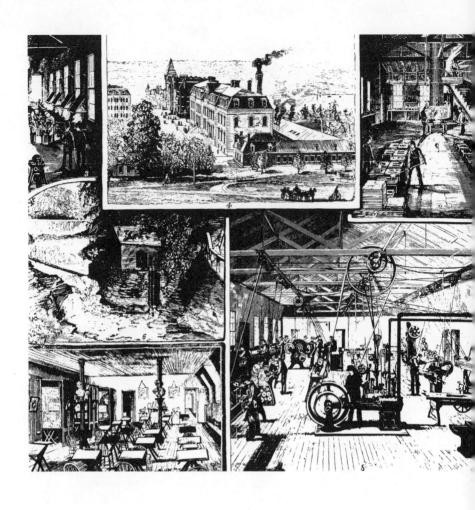

General view of buildings and machine shop at Sibley College of
Engineering, Cornell University, Ithaca, N. Y.,
1885 (*Scientific American*).

capable of giving expression to the opinions of the mechanical engineering ability of the country," and he hoped that the ASME's power would be used to help defeat the metric system. He had never met "a single representative of our plain, practical profession who advocated it out and out." Those who did, Worthington added, were "men who calculated eclipses and other things, with which I have very little to do."[32] Charles T. Porter, steam engine designer and manufacturer, seconded Worthington's remarks, and a general discussion began which ended only when it was decided that a letter ballot would be sent to all members so that they could vote on the issue. The results of the vote were announced at the Hartford meeting in 1881, and of 135 votes, 111 voted yes and only 24 voted no to the proposition: "The Society deprecates any legislation tending to make obligatory the introduction of the metric system of measurement into our industrial establishments."[33]

With the exception of occasional scuffles over the matter, the debate did not erupt in force until the New York meeting of the ASME in December, 1896, nearly fifteen years later, when it was revived by the delivery of a lengthy paper on "The Metric vs. the Duodecimal System" by George W. Colles. Reviewing with scholarly precision the whole history of the campaign for metric adoption, Colles then called upon the mechanical engineers present "to weigh the political, social and economic sides of questions." He maintained that the metric system was neither scientific, natural, nor uniform, and in addition, "it is the general verdict of engineers that the metric is *not* a convenient measure for the majority of purposes." He also raised a side issue, which was increasingly used by the opponents of the metric system, that it was a product of French or Latin impracticality whereas the foot and inch measure "was not the result of a few months' deliberation by a half-dozen noblemen who never even bought their own groceries."[34]

To be for metric adoption was, in Colles's view (as in that of Coleman Sellers), to advocate costly innovation with no hope of actual gain in efficiency. It was radical, and Colles suggested

---

[32] Comment on C. Sellers, "The Metric System," ASME, *Trans.*, I (1880), 25–26.
[33] "Proceedings of the Hartford Meeting, 1881," ASME, *Trans.*, II (1881), 9.
[34] ASME, *Trans.*, XVIII (1896–97), 493, 551, 553.

that it would be "socialistic" if the government were to adopt compulsory metric legislation. In short, to be for the metric system in the ASME was similar to being for Marx at a meeting of the chamber of commerce. The ASME had become and remained the major organized opposition to the adoption of the system in the United States. Dissension in the ranks was discouraged.[35]

Evidence of this can be read in the "confessions" of academicians who wished to remain in the favor of the shop culture elite. William Kent admitted that as a student and young teacher he had been enticed by the metric myth of uniformity, since the professors tried to cram it into him in the technical schools, but he had since recognized the error of his ways.[36] Another confessed to an early "flirtation" with the metric system and similarly denounced it as a fallacious system.[37] While remaining generally silent on the matter of his own opinions on the metric system, Robert H. Thurston acted on more than one occasion as the society's apologist for its antimetric stand.[38]

A few hardy individuals rose to the defense of the metric system after Colles gave his paper, attacking his assumption that the support of colleges and college professors by itself made the system impractical, and questioning whether the society really wished to publish in its *Transactions* Colles's description of France as a whimsical country; but the majority opinion was with Colles.[39] By the next meeting, the council had appointed a committee to prepare material which could be used in "opposition to legislation seeking to make the Metric System and its use compulsory in the United States." Its members were Coleman Sellers, Coleman Sellers, Jr., John E. Sweet, Charles T. Porter, and George M. Bond, all members of the shop culture elite.[40]

35 *Ibid.*, p. 590.

36 *Ibid.*, p. 594. Kent was not a teacher through most of his career, but had worked under Thurston as an instructor.

37 Leonard Waldo, comment on Colles, "Metric vs. Duodecimal," *ibid.*, p. 595.

38 "The Meaning of the Action of the A.S.M.E. on the Metric System," *Amer. Mach.*, December 18, 1902, pp. 1837–38.

39 A. L. Rohrer, comment on Colles, "Metric vs. Duodecimal," ASME, *Trans.*, XVIII (1896–97), 597–98.

40 "Proceedings of New York Meeting," ASME, *Trans.*, XVIII (1896–97), p. 10.

The matter was next brought to a head within the ASME when Fred Halsey gave a massive paper indicting the failure of the metric system (later elaborated on in a 1904 book, *The Metric Fallacy*) which was probably called forth by renewed attempts by metric supporters to persuade Congress to pass a bill making the system compulsory.[41] In December, 1902, Halsey delivered his paper and was promptly attacked by such men as M.I.T.'s president, Henry S. Pritchett, and Dr. A. E. Kennelly, and supported by George Colles, Samuel Webber, Asa M. Mattice (a designer with E. D. Leavitt's firm), Henry R. Towne, and (by letter) the grand old man of machine design, William Sellers. A few individuals didn't fit the pattern. For example, Arthur Falkenau (who, however, had come into contact extensively with other types of engineers) favored the metric system, as did Fred J. Miller, editor of the *American Machinist*. Nevertheless, the results of the letter ballot which followed the discussion bear out the assumption that this was a conflict of school and shop cultures.[42]

Amidst a series of denunciations of the system by the executive committee of the council of the ASME, a letter ballot was prepared and sent to all members during the spring of 1903 (see Table 5). On the question of the adoption of the metric system as the only legal standard in the United States, engineer-entrepreneurs (listed as businessmen and miscellaneous) voted five to one, mechanical and consulting engineers voted almost four to one, railroad men seven to one—against. Teachers, however, opposed it by a vote of only two to one, juniors voted about the same, and draftsmen were not even that heavily opposed. Thus, although all classes opposed it, those allied with entrepreneurial shop culture did so overwhelmingly, and those who were teachers of engineering and who were young (juniors and draftsmen) indicated much less dissent. This trend became even more clear in the vote on the question of legislation which

---

[41] Frederick A. Halsey, *The Metric Fallacy*, and Samuel S. Dale, *The Metric Failure in the Textile Industry* (two books published as one; New York: Van Nostrand, 1904).

[42] Discussion of F. A. Halsey, "The Metric System," ASME, *Trans.*, XXIV (1902–3), 489–522, 615–27. On Fred J. Miller's attitudes toward the metric question, see a copy of a letter from Miller to James Mapes Dodge, February 27, 1906, in Taylor Collection, and American Society of Mechanical Engineers, *Fred J. Miller* (New York: American Society of Mechanical Engineers, 1941), pp. 27–28.

would *promote* but not require the adoption of the metric system. The engineer-entrepreneurs still opposed the adoption by a vote of four to one, whereas mechanical and consulting engineers, draftsmen, and teachers split the vote evenly. On the question of whether metric adoption would actually be detrimental to the individual's "business," engineer-entrepreneurs and railroad men were still two to one against adoption on those grounds, whereas juniors were rather evenly split, with most feeling it was not detrimental; the majority of mechanical and consulting engineers felt no threat from the metric system. Engineering teachers by a vote of more than three to one showed that they felt no threat, and draftsmen were evenly divided on the question. These figures tell us clearly that teachers of engineering were the most favorably disposed to the metric innovation, and the shop elite were the most adamantly opposed. School and shop opposed each other on the issue of metric adoption, and it was primarily an issue within the context of bureaucracy versus entrepreneurship rather than a simple case of science versus backwardness.[43]

This strong difference of opinion between school and shop on the question of metric adoption reinforced the different attitudes which each culture had toward professionalization. The shop culture group was opposed because such adoption would cost them (as entrepreneurs) money without appreciable gain in efficiency. They were secure in their status and saw no need to make special efforts to curry favor with or placate the scientific community. Certainly they did not feel that metric adoption would in any way increase their status. The engineering educators, on the other hand, stood to lose no money by such an adoption. Although many of them were secure in their status, they did feel it necessary to do anything which might promote the welfare of their graduates. What could be more of a boost to the status of the college-trained engineer than for him to possess a new and arbitrarily determined system of measurement? What could better assure his ascendance over the boy from the shop? Knowledge of the metric system could

43 Table and analysis of metric vote, ASME, *Trans.*, XXIV (1902-3), 855-56.

TABLE 5: ASME Metric Vote Analyzed by Occupation in 1903

Vote	Foreign	Draftsmen	Railroad Men	Teachers	Mechanical and Consulting Engineers	Juniors	Business Men and Miscellaneous
In favor of the adoption of the metric system as the only legal standard in the United States	0	6	2	12	28	20	35
Opposed to the above	3	9	14	27	86	50	174
In favor of adoption of H.R. Bill No. 2054	0	5	0	15	25	20	30
Opposed to the above	1	9	14	27	82	46	163
In favor of legislation which would promote the adoption of the metric system	2	8	4	20	52	26	41
Opposed to the above	1	9	12	21	66	43	159
The substitution of the metric system for the English would be detrimental to my business	2	8	9	7	56	35	125
Would not	1	4	4	14	47	28	47
The substitution of the metric for the English system would be of advantage to my business	1	4	0	15	28	16	25

become, like calculus, a badge of the formally trained, which might increase the esteem with which the engineering community and the outside world regarded them. Metric adoption could destroy in a moment one of the few advantages which the shop man had over the technical graduate: the ability to make relatively complicated calculations accurately in seconds. It would cheapen this qualification, not so much by making the new system inaccessible to the shop man, but by making mental calculation a simple matter for the technical graduate.[44]

[44] See Hitchcock, *My 50 Years*, p. 62, for a description of the difficulty shop men had in adjusting to the introduction of the decimal division in the inch.

"*Too often the engineer tries to put his technical knowledge
and experience in a position of far greater value than it
deserves, for while it is important, business experience and
a knowledge of men and affairs are just as important
in order to fit properly a man for engineering
or manufacturing executive work.*"

Editorial
American Machinist, 1908

The ASME and the leading technical journals took little or no note of the efforts of the stationary engineer to achieve semi-professional status. Formed in 1882, only two years after the ASME, the National Association of Stationary Engineers (NASE) was a group devoted to getting uniform and operative license laws for stationary engineers, developing the moral character, social standing, and intellect of its members, and raising the general level and status of their occupation to at least semiprofessional if not full professional status. Some of the men occupied in running steam boilers and engines were mechanical engineers, even though stationary engineering was not a major contributor of personnel to mechanical engineering. Nevertheless, none of the officers of the NASE from 1882 to 1886 were members of the ASME, nor did the ASME ever take official notice of the existence and activities of the other organization.[1]

Stationary engineering was not, of course, entrepreneurial; the practitioner almost invariably was one salaried employee in a large, impersonal organization. The work of the stationary engineer was not of the type that would stimulate him to develop new concepts in the science and practice of engineering which he might present to a society such as the ASME. The NASE did exhibit certain attitudes which would have been likely to find favor with the members of the ASME, such as actively discouraging the members from taking part in labor movements. At the founding meeting of the NASE, the chairman warned the membership to stay away from "rum and trade unions" if the steam engineer were someday to be considered as something better than "grease slinger or coal heaver."[2]

Several other factors militated against any warm acceptance of the new organization by the ASME. The NASE, for example, engaged in active recruiting for members among the young

---

[1] See list of officers in *Amer. Mach.*, November 11, 1882, p. 8.

[2] *Ibid.*; "Steam Engineers and Strikes," *Engineer* (N.Y.), July 15, 1899, p. 164.

mechanics, machinists, semiengineers, and engineers in the shops. It thus presented itself as a rival organization to the ASME, in theory, even though the ASME would most probably have accepted only a few of the individuals forming the NASE. To further complicate the matter, the appeal for young machinists and engineers to join the NASE was presented as an antishop-culture argument. A. M. Davy of the NASE wrote the *American Machinist* in 1882 that the young mechanic must not say: "Mr. A or Mr. B are not good men because they don't know how to do machine work. I am a machinist; I learned my trade in the shop; I am not going to join a society that has for its object the educating of men to be engineers."[3] An organization which tried in its ideology to outprofessionalize the ASME, which in practice was composed of subprofessional engine drivers, was obviously not going to be taken seriously by the ASME or by individual mechanical engineers who were members of the shop culture elite. In fact, it seems that the elite studiously ignored the stationary engineers' organization altogether. The mechanical engineer saw no purpose in recognizing an organization whose membership might "work from 13 to 15 hours every day, and in the winter . . . is supposed to be early astir getting steam . . . after this . . . is expected to clean up the yard or the cellar, pack boxes . . . or anything else required."[4]

The same sins of omission were true of the ASME regarding other types of engine drivers, such as the marine engineers, whose organization and aims were quite similar to those of the NASE, and the locomotive operators or engineers, whose involvement in trade-unionism obviously made them pariahs to the professional mechanical engineer. Officers of the organizations representing these groups did not get into the ASME. Several things are common to all these groups: they used the title "engineer," they operated steam engines, and they had no relationship to shop culture. While mechanical engineers as individuals might have personal, social, and occupational relationships with engine and boiler operators on various occasions,

3 Letter to the Editor, *Amer. Mach.*, December 23, 1882, p. 6.
4 H. L. Stellwagen, Letter to the Editor, *Amer. Mach.*, May 29, 1880, p. 6.

the mechanical engineer as professional had nothing to do with them.

The case of the shop machinist, who in the minds of many had inherited the untarnished mantle of the mechanic, was different indeed. In the meetings of the ASME, on the pages of the leading technical journals, and everywhere mechanical engineers met and talked, the status and role of the mechanic-machinist was a subject of concern, and the interrelationship between engineer and machinist was under constant and watchful scrutiny. The shop elite had an interest in preserving the structural relationships they believed to exist in the shop between the man at the bench and the man in the office.

One basic question was whether there would be a place in the shop of the future for the versatile, ingenious mechanic of the past. Would the machinist (and this term was increasingly being recognized as the correct designation) continue as a source of personnel for the profession of mechanical engineering or would he be made into an unthinking robot by automatic machine tools? The verdict of the leading technical journals and of the shop elite was the former. *American Machinist* suggested in 1882 that the machinist's position "in relation to improved machine production, and the tools and appliances of the future, however refined, can never usurp his place in the industries of the world, nor relieve him of the part he is to play in future mechanical advancement."[5] The difference, *Mechanical Engineer* proposed in 1881, was that which exists between a workman and a mechanic: "A workman is a man who works, and he may be a laborer in some instances, but a mechanic is a man who devises."[6] There would always be, according to both journals, a need for the skilled mechanic-machinist, and he would not be forced into running automatic machines that produced the same part several hundred times a day.

Implicit in this view of the machinist was the idea that the ultimate goal of any self-respecting mechanic or machinist was to become an entrepreneur. *American Machinist* suggested that

---

5 January 21, 1882, p. 3.
6 "Workman and Mechanic," *Mechanical Engineer*, February 12, 1881, p. 53.

every really successful mechanic had to have some executive ability,[7] and *The Northwestern Mechanic* went even further: "Is there a mechanic worthy of the name who has not the hope of some day owning a shop of his own?"[8] Secondly, this conception made a pariah of the tramp mechanic, a common figure in the technical journals of the 1880's. Technical journals almost unanimously discouraged young men from becoming involved with the roving, footloose, and fancy-free machinist who invariably lacked the virtues of sobriety and stability. One article flatly stated that "machinists who remain the longest in one shop are the most successful in accumulating means to start business on their own account."[9] The tramp mechanic was, at least according to the mythology built up by the technical press, also the individual most likely to be involved in fomenting labor troubles. "The Grumbler" described in an 1880 article in the *American Machinist* was this sort of person: "A strike takes place in a certain shop in an adjoining town. Grumbler goes for all the items concerning the same, and returns to the shop with the astounding announcement that the strike is but in its infancy and will spread all over the country. . . . We'll show them whether we have any rights or not." The journal warned the young machinist, "Never listen to Grumblers, and, above all things, avoid being influenced by what they say."[10]

Benevolent and mutual aid societies were approved of, but associations that attempted to invade the shop and to separate workmen, foremen, and engineer-proprietor were not. The machinist who joined a union was obviously no longer fit to be considered a potential engineer. This pattern of thought was so frequently followed by the journals that a St. Louis paper, *The Carpenter*, published by an association of cooperative trade-unions, referred in 1881 to "*The American Machinist* and other capitalistic journals."[11]

Specialization was, of course, one of the things that had led

7 "Executive Ability among Mechanics," *Amer. Mach.*, March 17, 1892, p. 8.

8 "Two Ways of Beginning," *The Northwestern Mechanic, A Journal for Machinists, Engineers, Founders, Boiler Makers, Blacksmiths and Miners* (hereafter cited as *Northwestern Mechanic*), I (March, 1889), 8.

9 "Migratory Machinists," *Amer. Mach.*, May 29, 1880, p. 8.

10 March 27, 1880, p. 8.

11 Quoted in *Amer. Mach.*, July 16, 1881, p. 9.

to the routinization of the work of the machinist and ultimately to trade-unions, and most mechanical engineers were against it when they thought in terms of machinist or mechanic as an occupation leading into engineering (a different attitude might and did result when they thought in terms of production problems and getting the work done). *Scientific Machinist* told a story about Uncle Dan who was a machinist and Joe who was a specialist. "Uncle Dan could take the machine shop to the mill; Joe could not see any way but to take the mill to the machine shop. Uncle Dan would do a good job on any tool in the shop, and he could do a good job 50 miles away from these tools. Joe couldn't do anything away from the planers and the shaper."[12] *Practical Mechanic* also extolled the virtues of the jack-of-all-trades mechanic, but bemoaned the fact that most men were specialists.[13] *Northwestern Mechanic* noted that one need only look at the help-wanted section of the newspaper to see that it was specialists who were actually in demand, not the general mechanic.[14] An article in the University of Minnesota engineering school magazine suggested that the mechanic as well as the engineer would have to specialize to succeed in the twentieth century.[15]

For the mechanic to rise into mechanical engineering he would need, in addition to a general facility in the shop, instruction in mathematics and other elements of engineering science. One great question was how to accomplish this. We have already seen how much interest the leading engineers exhibited in trade, technical, and industrial schools of all types. Their approval of the Worcester Free Institute and its program is instructive, since there the machinist and mechanical engineer were on a continuum and the lines between them blurred. Technical journals exhorted the young machinist to learn drafting in night school, discussed various plans for shop schools, asked the textbook and handbook writers for simple, clear compila-

12 F. F. Hemenway, "About Apprentices," *Scientific Machinist*, May 15, 1896, p. 11.
13 F. J. Masten, "Skilled Workmen," quoting an earlier editorial, *Practical Mechanic*, II (July, 1888), 6.
14 "The Mechanic—Skilled and Unskilled," *Northwestern Mechanic*, II (November, 1890), 8.
15 O. P. Briggs, "The Application of a Technical Education," Minn. Engr., *Year Book*, X (1902), 50.

tions of engineering practice and science, and debated the merits and faults of the apprentice system.

On rare occasions the technical editors admitted that the problem was not one that could be solved by exhortation, but was one which arose "from the changed systems of doing mechanical work. Only a part of the men who work in machine shops are machinists." What were those who were not? *Scientific Machinist* offered the answer: "We have noticed that where young men are serving an apprenticeship, they are not being taught like they used to be. They are simply put on the lathe, the drill press, the planing machine, and perhaps a few more jobs, and kept at very much the same line of work for months at a time so that when they get through, they are expert machine operators, but not expert machinists as that name has been understood."[16] When the mechanical engineer tried to educate such relatively unskilled and unimaginative workmen in the time-honored tradition of shop culture, he was often met with the words "What the d– – –l are logarithms?" and a clear indication that the individual asking the question had no interest in finding out.[17]

The *American Machinist* in its early years exhibited some interest in the problem of the social status of the mechanic and machinist. Quirk, a pen name for an anonymous contributor, wrote with indignation of "young clerks" being invited to attend a church picnic, an occasion which brought no mention of "mechanics." He hoped that the ASME would be able to lend stature to the mechanic by its very existence and character.[18] Generally, however, it was the opinion of mechanical periodicals that one got just about as much status as one earned, and there was remarkably little sympathy given to mechanics or machinists who complained of being "snubbed" by polite society. Frequently, wrote the editors of *American Machinist*, those who feel this way "have been mistaken in their estimate of what constitutes society and have been seeking admission

16 "The Shop School, Our Only Hope," *Scientific Machinist*, May 15, 1896, p. 4.
17 "The Average Mechanic and Education," *Amer. Mach.*, May 8, 1880, pp. 8–9.
18 Quirk, "Social Position," *Amer. Mach.*, June 11, 1881, p. 7.

into showy circles in which really good society would find little of either pleasure or profit."[19]

If there was a possibility for relationship and for upward mobility from machinist to mechanical engineer, it was according to the rules laid down by the shop culture elite and the technical journals. These rules demanded that the mechanics or machinists assume the character attributes of the mechanical engineer from the outset. They had to be sober,[20] skilled, versatile, and innovative, yet also respectful of superiors; they had to act as though they would someday be entrepreneurs and shop executives themselves; and they had to avoid contact with any sort of trade-union agitation. These were the conditions for a good relationship—they were seldom met.

Many of the shops which employed mechanical engineers and machinists no longer practiced true apprenticeship. What had once been small, experimental shops had often become large manufacturing plants by the 1890's, and there was no time for training machinists broadly or for trying to teach them engineering. It was better, as the shop became a factory, to put a man at one machine and to keep him there if he learned his job well. The new relationship that resulted from these changes in the nature of the shop was part of the well-spring of Taylorism and scientific management. Both the change and the response made the shop culture mobility ideal increasingly a myth.

[19] "A Plea for Machinists, from a Machinist's Sister," *Amer. Mach.*, April 8, 1882, pp. 8–9.

[20] Regarding the sobriety expected of both mechanic and engineer alike, see A. L. Holley to John C. Thompson, June 4, 1877, in Holley Letterbooks.

CHAPTER **11** ENGINEERING: THE FRAGMENTED
PROFESSION

There were distinct differences between types of professional engineers, based sometimes on structural differences in their respective employment roles, sometimes on the extent to which they had a nucleus (such as the shop) for development of a professional culture outside the professional association and the school. Other factors that caused differences were geographical distribution and mobility, competition between professional associations, and participation in local, state, and national government. The existence and particularly the activities of the four "founder" societies, the ASCE, AIME, ASME, and American Institute of Electrical Engineers (AIEE), led to ideological conflicts between engineers as professionals that often were not represented by actual conflict in practice. As we have seen, different types of engineers did in fact get along quite well and find many areas of mutual agreement and action in local and regional engineering societies and clubs. It seems likely that much of the conflict between types of engineers existed only at the national professional association level and was the product of a few ideologists. This ideological conflict was strong enough, however, to prevent the formation of a truly inclusive national professional organization for all engineers. The reluctant keystone in the arch of engineering unity was the American Society of Civil Engineers.

The American civil engineer—through his professional association after 1867, and often individually before and after that date—regarded civil engineering as a broad profession including all types of engineers; he resented efforts to divide the field into specialties. ASCE president William J. McAlpine stressed in his inaugural address in 1868 that the "American Society of Engineers," as he chose to call it, included all types of engineers, the two major subdivisions at that time being mechanical and mining.[1] This position was not un-

---

[1] American Society of Civil Engineers, *Transactions* (hereafter cited as ASCE, ·*Trans.*), I (1867-71), 55.

New erecting shop for the Gaskill horizontal compound pumping
engine, Holly Manufacturing Company, Lockport,
N.Y., 1888 (*Scientific American*).

justified since the European tradition which had originally differentiated civil and military engineers included all the former within "engineering," even though they might be specialists. The first major blow to this position came a mere four years after the civil engineers' organization in the United States had revived itself.

The American Institute of Mining Engineers was founded in 1871 and was in many respects a clear challenge to the inclusiveness of the ASCE. The injury to the civil engineers' organization was felt so deeply that for a time they refused to have anything to do with the new group. The realities of the situation seem to have mollified this anger by 1880, when the ASME was founded, since the civil engineers' organization was helpful in giving advice to the mechanical engineers and even provided them with a place to meet in their early years. Occasionally, throughout the period from 1871 to the mid-nineties, presidents and members of the ASCE expressed themselves on the question of inclusiveness, but the issue was not a live one. W. Milnor Roberts in 1873 and Max Becker in 1889 spoke on the subject in their inaugural addresses, but Becker was almost conciliatory, suggesting that specialization was necessary and that civil engineers simply had to accept it.[2]

The technical journals, particularly *Engineering News*, the civil engineers' equivalent of the *American Machinist*, kept the issue alive over the years, but the ASCE chose to keep it out of its discussions and formal transactions. About 1895, spurred perhaps by talk of engineering unity at the Chicago Exposition in 1893, the ASCE began to show more interest in the issue. George S. Morison made it a major topic in his presidential address before the society in that year, and it was a major concern of almost every presidential address from that year until 1910. Morison suggested that "we are Priests of the new

---

[2] Max J. Becker, "Address at the Annual Convention, 1889," ASCE, *Trans.*, XX (1889), 234; W. Milnor Roberts, "Engineering," ASCE, *Trans.*, II (1873), 69.

In response to a letter from A. L. Holley on the prospective founding of the ASME, William Rich Hutton, prominent civil engineer, replied that the creation of additional engineering societies was unwise in that it would weaken the existing ones. Dissatisfaction with the ASCE should be met, not by founding new societies, but by working to improve that organization. Hutton to Holley, April 2, 1880, in William Rich Hutton Papers, Division of Mechanical and Civil Engineering, Smithsonian Institution, Washington, D.C.

epoch, without superstitions," implying a special master role for the civil engineer.[3] The following year President Thomas Curtis Clark made an even stronger case for the civil engineer as master engineer: "The civil engineer's true position is similar to that of the architect, who commands the services of many different professions and handicrafts. Mechanical and electrical engineers and many others aid him in their several lines, but from him comes the comprehensive design, and he alone is the director-general of the works." Clark dismissed lightly the typical division of college curricula, which suggested that civil engineers were concerned only with structural engineering, and recognized mechanical, mining, and electrical as types, not branches, of engineering.[4] In 1897 B. M. Harrod, speaking at the opening of the society's large, new house in New York, talked of "mechanical, sanitary, the electrical and other classes of civil engineers."[5]

The ASCE attempted throughout the period from 1871 to 1910 to justify a claim to inclusiveness by admitting to membership a very few prominent practitioners of the other specialties. The claim was not justified, however, in the matter of presentation and publication of papers dealing with the subject specialties. Before the formation of the other founder societies, the ASCE was a forum for such presentation. Before 1871 and even up to about 1875, papers were given on mining and on mechanical engineering topics, but few were offered in the area of electrical engineering. By 1895, however, when the stress on inclusiveness really became a major issue in the ASCE, the only topics presented at meetings were those dealing with structural engineering. It was, ironically, just when the civil engineers' organization no longer spoke for the engineering profession as a whole that it became the most interested in doing so.

A good example of the ASCE's promotion of this idea in the early twentieth century was the International Engineering Congress which the society sponsored and organized for the St. Louis Fair of 1904. The ASCE appeared to recognize specialization in engineering by having numerous sections on such topics

3 ASCE, *Trans.*, XXXIII (1895), 483.
4 "Science and Engineering," ASCE, *Trans.*, XXXV (1896), 517.
5 "Address, November 24, 1897," ASCE., *Trans.*, XXXVIII (1897), 425.

as mechanical and electrical, but leaving out, conspicuously, mining engineering. However, a check of the individuals who gave papers at the meetings reveals that almost none of those in mechanical engineering, for example, were at all prominent in the ASME and many were not even members. They were members of the ASCE who practiced mechanical engineering, but who had only tenuous connections with the mechanical engineering professional association. The same was true for the electrical engineers. The congress gave the impression that the other professional associations had made no contribution to knowledge, and it no doubt convinced and influenced many laymen.[6]

Some civil engineers began to tire of this game after ten years, and by 1906 one can recognize the beginning of a countermovement in the ASCE leading toward a more realistic relationship formally and informally with other kinds of engineers. There was after that date a tendency for the other branches of engineering to be openly referred to in meetings as free and independent specialties. This coincided with the opening of a million-dollar union engineering headquarters for the other three societies and represented a fear, no doubt, on the part of some civil engineers that their extreme position could lead to a great loss of importance for their organization. By 1909 President Onward Bates, while giving lip service to the ideal of inclusiveness, admitted that many of the men on a list of civil engineers he had assembled "are also members of the National Societies of Mechanical, Electrical, and Mining Engineers, who, if restricted to one Society, might not elect to be classed among the Civil Engineers."[7] Although the issue remained a source of irritation and conflict between civil engineers, mechanical engineers, and others until about the time of American participation in the First World War, by 1910 it was no longer an issue as far as the whole profession of engineering was concerned.

Another indication that the period 1905 to 1910 heralded a

6 List of Participants at International Engineering Congress, ASCE, *Trans.*, LIV (1905), p. 466. The ASME had been asked to participate, but had declined. ASME Council Minutes, November 6, 1903.

7 Onward Bates, "Address at 41st Annual Convention, Bretton Woods, N.H., July 6, 1909," ASCE, *Trans.*, LXIV (1909), 571.

changing attitude on the part of civil engineers regarding mechanical engineers can be seen in curriculum changes at Rensselaer Polytechnic Institute in Troy, New York. That institution had long been the elite training school for professional civil engineering and had concentrated on that one specialty. Although courses in mechanical and other types of engineering were theoretically set up in the mid-sixties, they were unsuccessful and were eliminated in a curriculum reorganization in 1870 when it was decided to limit the school to training in civil engineering. Ironically, Alexander Lyman Holley was on the committee that made this recommendation in 1870, a full ten years before he helped to found the ASME.[8] By the nineties, like the professional association and the technical press, Rensselaer was riding hard the idea of civil engineering inclusiveness and refused to think of adding a special course in mechanical engineering. Washington A. Roebling, a civil engineer himself, complained of the narrowness of training at Rensselaer in a letter to Robert H. Thurston in 1887: "I am temporarily residing at Troy, where my son is attending the Institute.

"I note that quite a number of the graduates go to Cornell for a few years afterwards to put on the finishing touches. There ought to be a little more practical Mechanics at Troy, it is rather too exclusively devoted to Engineering and Mathematics."[9]

There was a rising demand that Rensselaer diversify, and the issue came to a head in 1907 when the entire faculty voted to establish schools of mechanical engineering and electrical engineering. The debate had begun intensively in 1906 and had included graduates, faculty, and trustees. One demand of the inclusiveness school, which was accepted to some extent, was that even if the M.E. and E.E. programs were added, those graduating from them should still be trained as versatile engineers able to work in any branch of the profession. In spite of the 1907 capitulation the school retained its image, and by 1914 only thirty-eight individuals had gone through the mechanical engineering program.[10]

8 Ricketts, History, pp. 136, 189.
9 August 7, 1887, in Thurston Papers.
10 Ricketts, History, p. 139.

In general, relations between the civil engineers as professionals and the mechanical engineers were better than those between the civil engineers and the other two specialties. The ASME alone of the three other societies had occasion to thank the ASCE officially for its "advice, assistance, and encouragement."[11] For the first several years of the ASME's existence, the ASCE even loaned a room in their house to the ASME for its board meetings, a favor extended to neither of the other two societies. Nevertheless, shortly after 1880, papers on mechanical engineering topics were no longer presented to the ASCE, mechanical engineering accomplishments were dropped from the annual review of achievements in engineering, and occasionally aspersions were cast upon the professional character of mechanical engineers. Another example of lack of communication was a historical talk on the beginnings of engineering given before the ASCE in 1891. In dealing with the achievements of mechanical engineering, the survey did not get beyond Oliver Evans and John Ericsson, revealing a very unrealistic and hopelessly outdated idea of the scope and accomplishments of mechanical engineering. The same author in mentioning "Engineering Societies" and "Engineering Schools" mentioned only those which were predominantly for the benefit of civil engineers.[12]

In 1895 the inclusiveness argument was applied with equal vigor to the mechanical engineers. As George S. Morison expressed it, "Any man who is thoroughly capable of understanding and handling a machine may be called a mechanical engineer, but only he who knows the principle behind that machine so thoroughly that he would be able to design it or to adapt it to a new purpose . . . can be classed as a civil engineer." He accused the ASME of taking in mere superintendents and businessmen, which was all right for them, but not acceptable for a professional association of civil engineers.[13]

Civil engineers always had a higher percentage of consultants among their ranks, and this role permitted many to assume

11 "Proceedings, December," ASME, *Trans.*, II (1881), 421.

12 J. Elfreth Watkins, "The Beginnings of Engineering," ASCE, *Trans.*, XXIV (1891), 356–60, 378–89.

13 "Address at the Annual Convention at the Hotel Pemberton, Hull, Mass., June 19th, 1895," ASCE, *Trans.*, XXXIII (1895), 471–72, 477.

holier-than-thou positions regarding the issues of professionalization and relations with other branches of engineering, including mechanical. The frequency of the consultant role with
its attendant freedom from employer control may never have
been as high as the civil engineers believed it to be, although
concrete figures on this are lacking for the early period. By
1909, however, a study by the ASCE revealed that three-
quarters of their membership worked for salaries on a regular
year-round basis for one employer.[14] Even if we assume that
only one-half of the practicing civil engineers in 1850 were
consultants, this would still mean that a large minority through
the 1850-to-1910 period were consultants or entrepreneurs and
could have had a very large voice in the direction of the ASCE.
It seems unlikely that large percentages of civil engineers were
entrepreneurs, since the nature of their work (building roads,
bridges, aqueducts, railroad beds, and so on) led to work for
either governmental institutions or large bureaucratic corporations. This inference leads to the conclusion that as many as
one-third of the civil engineers in the ASCE (admittedly the
elite) were consultants in 1895. The percentage was probably
dropping, and President Octave Chanute told the ASCE in 1891
that only as more civil engineers became consultants would
they see "a marked improvement in the independence, in the
standing, and in the emoluments of the civil engineers."[15]

Some of the emphasis on professional role came from the
painful and difficult early years of civil engineering in this
country. In order to survive, as a profession and as individuals,
in the entrepreneurial setting of canal and railroad development
from 1810 to 1860, American civil engineers had made a special
effort to establish the role of the engineer in relation to a
project as a distinct and different approach from that of a
financial partner or entrepreneur. In a way not common to the
engineering specialties which developed in the post-Civil War
era, the civil engineer had made a special point of the fact that

14 Bates, "Address at 41st Annual Convention," pp. 571–72.
15 "Address at the Annual Convention at Chattanooga, Tenn., May 22.D, 1891," ASCE,
Trans., XXIV (1891), 429.

he was not an entrepreneur. This is reflected in the type of professional culture he developed.[16]

If the American civil engineer can be said to have had anything comparable to the shop culture of the mechanical engineer, it would be something called "field culture." It was, in effect, a group of men in the field surveying a road, a bridge, or other project, composed of every level and stage of development in the profession, from rod-man, to surveyor, to the chief engineer of the project. It was felt by many civil engineers that this association in the field had a socializing and educating function which would compare with the functions expected of shop culture by the mechanical engineers. Obviously, field culture was not entrepreneurial; shop culture was. The orientation of field culture toward the profit system was somewhat oblique and certainly not ever-present.

In this context, civil engineers differed from mechanical engineers in that they often expressed distinct hostility toward the "financial interests" and dissatisfaction with the working of the American capitalistic system. Octave Chanute told the members of the ASCE in 1880 that to reach higher professional ground they must stop allowing themselves to be "ordered hither and thither by promoters of schemes, and the magnates of Wall Street."[17] George F. Swain, president of the Boston Society of Civil Engineers, told that organization in 1897 that the credit for great engineering undertakings was often mistakenly given to "some financier or promoter."[18] There is almost the rejection of urban-industrial society in the complaint of ASCE president Ashbel Welch in 1882 that "nothing is done now by the individual, but everything by some institution, or corporation, or central power, or great firm. Man has ceased to be a unit, and become only an atom of a mass."[19] One would look long and hard to find mechanical engineers expressing such sentiments.

16 Calhoun, *American Civil Engineer.*
17 "Annual Address, 1880," ASCE, *Trans.*, IX (1880), 254.
18 "The Status of the Engineer," AES, *Jour.*, XVIII (1897), 183.
19 "Address of Ashbel Welch, President, A.S.C.E.," ASCE, *Trans.*, XI (1882), 156.

This antientrepreneurial bias, combined with the fact that many civil engineers in the post-1860 era were municipal engineers, led to considerable interest in city planning and urban reform that was to be accomplished under a sort of technocratic system manned by engineer-administrators. Benjamin M. Harrod, ASCE president in 1897, told the society that "in cities and in many communities the duties of the government rest more upon good engineering than on legal skill." The leaders of the future city and state would be civil engineers, he prophesied, by virtue of their special training in the "utilitarian powers of nature." Here he stressed that civil engineering was not a "mechanical art" because of its service aspect. Civil—not mechanical—engineers were the true professionals, according to Harrod.[20]

Many civil engineers, both those permanently employed under salaries and those called in on a temporary consultant basis, developed severe role conflicts in which they were forced to choose between their professional ethical position and a less ethical position offered by an employer. Particularly common were cases where the engineer in charge of a public work of some kind was interfered with and subordinated to corrupt political and business interests. In 1860, in such an incident involving the building of the Croton Aqueduct in New York, Alfred W. Craven, the chief engineer, resented interference with his work by the mayor of the city, and charged: "I am an engineer, charged with the responsibility of the plan and execution of a public work requiring skill and science. He is a layman, and law is not guilty of the absurdity of making me 'subordinate' to him in the discharge of my duty."[21]

The different role played by civil engineers led to greater interest in questions of status than the mechanical engineer exhibited. Sometimes these concerns were merely more extreme versions of positions held by professional agitators among the mechanical engineers, such as interest in title regulation and in whether technical schools were to give the professional title or

---

20 "Address at the Annual Convention at the Chateau Frontenac, Quebec, June 30th, 1897," ASCE, *Trans.*, XXXVII (1897), 539–41.

21 "The Mayor of New York and the Croton Aqueduct Engineer," *Engineer* (Phila.), September 13, 1860, p. 35.

merely a bachelor's degree, concern over acceptance or non-acceptance of expert testimony in courts of law, the gaining of professional status similar to that held by doctors and lawyers, and the improvement of engineering schools. Frequently, however, the status concern of civil engineers went beyond the areas which were explored by the mechanical engineers. For example, considerable time was spent in the meetings of the ASCE discussing the actual possibilities of setting up standards for determining who should be allowed to practice civil engineering. There was disagreement over whether the national government, the state governments, the society, or the school should set and administer these standards, but many civil engineers felt that the right to practice should be regulated in a systematic way and quackery eliminated.[22]

The high percentage of consultants and the independent professional image held by many civil engineers led to a concern for an action to create a uniform code of ethics for the profession, a subject the mechanical engineers did not become much concerned about. A whole series of papers were given before the ASCE on the problems of relations between the engineer and those with whom he came into professional contact, on such related topics as the ethics of advertising one's services, owning and exploiting patents, relations with other kinds of engineers, passing formal judgments on brother engineers, relation of private life to professional standing, the practicability of the ASCE exposing quacks publicly, and what acts made the engineer liable to be censured by the ASCE. The questions asked ran the gamut of professional behavior in an ethical framework.[23]

After 1893 the question of a code of ethics became a perennial issue, and although the final votes in the nineties rejected

22 "Civil Engineering Practice: Shall It Be Regulated by Law," an informal discussion, ASCE, *Trans.*, XLVI (1901), 129–40; Robert E. McMath, "Engineers, Their Relation and Standing," an address before the Engineer's Club of St. Louis, AES, *Jour.*, VI (1886–87), 94–100.

23 "The Relation of the Engineer to Those with Whom He Comes in Professional Contact," a series of papers read before the Boston Society of Engineers, AES, *Jour.*, XII (1893), 437–53; Archibald R. Eldridge, "Is It Unprofessional for an Engineer to Be a Patentee?" ASCE, *Trans.*, XLVIII (1902), 314–17; "The Regulation of Engineering Practice By a Code of Ethics," an informal discussion, ASCE, *Trans.*, XLIX (1902), 45–62; Albert J. Himes, "The Position of the Constructing Engineer and His Duties in Relation to Inspection and the Enforcement of Contracts," ASCE, *Trans.*, LVI (1906), 117–19.

a code as unnecessary to professional status, by the early twentieth century the tide was turning the other way. The civil engineering technical press took up the issue in 1902, and by 1910 the division among civil engineers on this question was sharp. Though not resolved in this period, the question of a code of ethics was dealt with in depth by the civil engineers; this was not true of their mechanical comrades.

Civil engineers, at least those acting through the American Society of Civil Engineers, exhibited in this period greater professional orientation in the subjects they considered important for discussion than did any other types of engineers. Their firm attachment to the ideal of the independent consultant, coupled with a probably greater percentage of independent professional practitioners, led them occasionally to respond to the industrial world in a belligerent fashion. In this they often admired, but did not approach, the independent position of the architect in society. Mechanical engineers often joined with civil engineers in envy of this group, which they believed to be far more successfully professionalized than themselves.

We have seen how the mechanical engineers in particular reacted to what they thought to be great deference to and superior compensation for architects in connection with the Chicago Fair in 1893. The issue was the belief, held by many mechanical engineers, that the architect, because he was unscientific and unheedful of basic engineering principles, did not deserve the superior compensation and status they believed him to have. Such conflicts arose more frequently in the twentieth century and in large cities like New York, in connection with the design, construction, outfitting, and ornamentation of the skyscraper.

The multi-story, reinforced steel and concrete office building, which became a standard method of beating the high cost of urban land, required for its successful erection the combined services of architect, civil engineer, and mechanical engineer. Many engineers, both civil and mechanical, felt that the architect was only marginally necessary to the operation and some went so far as to regard him as a parasite. An editorial in the *American Railroad Journal* noted as early as 1856 that the architect was more frequently the artist than the engineer, and that "taste" in building often was synonomous with stupidity.[24] By the 1890's the issue was becoming a real one and the conflict of engineering and architectural purposes was becoming acute. One engineer told the members of the Technical Society of the Pacific Coast that historically the best styles of architecture had been those which reflected the strengths and characteristics of the materials used. In 1899, he maintained that the modern steel office building would have to develop a style determined by science, not art, if it was to remain within this tradition.[25] Leicester Allen, a mechanical engineer, wrote an article for *Engineering Magazine* in 1893 indicating the lack of cooperation between engineer and architect. Progress would take place in skyscraper design when the architect was stimulated by the placement of such engineering appliances as elevators, steam plants, and so on, that would structure the building rather than force him to hide them with inappropriate ornament.[26] The relationship between architect and engineer, although of concern to the few involved in it, never reached the forum of the ASME in any major way.

Civil engineering, much more than mechanical, was concerned with defining and delimiting professional behavior. This was primarily because of differences in the types of employment role they played. Many more civil than mechanical engineers were consultants or independent practitioners. As a result, civil engineers tended to think that mechanical engineers were mostly businessmen masquerading as engineers and that they were not interested in becoming truly professional. Relations between the two specialties on the level of professional organizations were cordial but never productive. For most practical purposes, civil and mechanical engineering were completely separate professions, with different views of what professional behavior ought to be.

When the American Institute of Mining and Metallurgical

24 "Qualifications for Architects," *Amer. Rr., Jour.*, April 5, 1856, p. 216.

25 G. W. Percy, "Mechanical Influence in Architecture," AES, *Jour.*, XXII (1899), 113–38.

26 "The Field of Domestic Engineering," *Engineering Magazine* (N.Y.), VI (1893–94), 87.

Engineers was formed in 1871, it was not a strong national organization and for some time its survival was doubtful. At a critical juncture in 1875, a number of fairly prominent engineers, who were by role mechanical, but who had continued to call themselves civil engineers, joined the new organization. Robert Thurston and Alex Holley were among this group which gave the AIME the stature it needed to survive. This migration of a number of mechanical engineers, attracted, no doubt, in part by the term "metallurgical" in the organization's formal and complete title, resulted in a good working relationship between the AIME and the mechanical engineers' organization formed in 1880. The group that migrated to the AIME in 1875 (presumably because these men could not find a forum for their ideas) was the same group that formed the ASME and provided that organization with some of its prime movers in the early years.[27]

One can only surmise that one of the reasons for the formation of the ASME so soon after the migration of mechanical engineers into the ranks of the AIME was the low standards for admission to membership adopted by the latter. The civil engineers were fond of remarking that the mining engineers would admit virtually anyone, and this was not far from the truth. The "member" category could include those "engaged in mining," and associate members need only be "persons desirous of being connected with the Institute." Mining engineering was very heavily entrepreneurial, and any kind of set, rigid standards would have excluded from membership many self-made, uneducated mine operators. While this would have been one possible mode of action, it was not the one chosen by the men who originally founded the AIME.[28] The AIME's golden years were from 1875 to 1880, when it contained simultaneously the elite of mining and mechanical engineering. Robert H. Thurston chose to present some of his best papers before the AIME in the late seventies, but stopped doing so altogether when the ASME was formed. There remained, however, much

27 R. W. Raymond, "Biographical Notice of Robert Henry Thurston," American Institute of Mining Engineers, *Transactions* (hereafter cited as AIME, *Trans.*), XXXV (1904), 428.

28 "Rules," AIME, *Trans.*, I (1871–73), xvii.

similarity of subject matter shared by the two branches of engineering, and there was no objection to the same man holding (at different times) the presidency and other top administrative jobs in both societies. This did not happen in the ASCE, which remained aloof from both in the matter of electing officers.

There were great differences in the conditions and environment of employment experienced by practicing mining and mechanical engineers. Mechanical engineers tended to be located in the urban, industrial Northeast, but mining engineers tended to be located, at least in the early stages of their careers, in the rugged mountains of the West, or even in South America or Africa. The most obvious exceptions were those connected with coal mining, most of whom, before 1910, were located in places like western Pennsylvania, not far from urban centers. This peculiarity was recognized and meetings of the AIME were held three to five times per year all over the country. The civil engineers had experienced difficulty in organizing for effective joint action when they, too, were spread out in the wilderness in the 1830-to-1865 era. The civil engineers had begun to return to the city, however, when it began to grow at a rapid rate, to solve the problems of mass transit, water supply, and the building of bridges. The mining engineers' geographic dislocation, combined with their low entrance requirements, effectively precluded their becoming a serious partner or competitor of the other societies.[29]

The geographical split within the AIME was also topical, with the subject of mining proper dominating the contributions of the western members and the problems of iron and steel technology occupying the interest of the eastern members. The society seems to have been run by an eastern establishment similar to the one which ran the ASME.[30] The AIME also shared with the younger ASME an orientation toward a culture (in this case a confused mixture of furnace, mill, and mine culture) which was opposed to the complete usurpation of the professionalizing process of the school. As a consequence, some

[29] AIME, *Trans.*, V (1876–77), 3, 29, and *passim;* "List of Members and Associates Arranged According to States and Towns," AIME, *Trans.*, XV (1886–87), xlii–li.

[30] William B. Potter, "Some Thoughts Relating to the American Institute of Mining Engineers and Its Mission," AIME, *Trans.*, XVII (1888–89), 489–94.

prominent mining engineers and entrepreneurs engaged in elaborate forms of paternalism to offer their subprofessional employees a chance to climb the ladder to the engineer's spot. Eckley B. Coxe, a president of both the AIME and the ASME, operated at his own expense a school for his employees at his mining operation at Drifton, Pennsylvania. He did this to establish a link between the workman and the school-trained engineer and simultaneously to eliminate the sources of socialism, communism, nihilism, internationals, and trade-unions.[31]

Mining engineers, unlike other types, showed a great interest in the correspondence school movement, which offered a solution to the problem of educational mobility for the miner or gang foreman of ability who aspired to be an engineer or at least a superintendent, but who was trapped in a mining town where no schools existed. Several of the correspondence schools, such as the giant International Correspondence School in Scranton, Pennsylvania, were set up in response to this kind of demand.[32]

Paternalistic concern for the physical conditions of the mine led to attitudes different from those of the mechanical engineers' ideas about the shop. Professor William P. Blake of Yale University told the AIME that the United States was behind Europe in providing such elemental necessities as decent housing, environments in mining towns suitable for families, and unadulterated beer and liquor sold to miners at cost.[33] R. W. Raymond presented an appeal to the organization to take the lead in improving mine hygiene, giving attention to such items as ventilation, temperature, interior transportation, and proper clothing.[34] President James C. Bayles (also prominent in the ASME, editor of *Iron Age*, and one-time Health Commissioner of New York City) gave a unique address to the AIME. Bayles felt that the organization was wrong in its belief that socialism and related evils could be eliminated by a liberal dash of

31 E. B. Coxe, "Secondary Technical Education," AIME, *Trans.*, VII (1878-79), 221.

32 H. H. Stoek, "The International Correspondence Schools, Scranton, Pa., with Special Reference to the Courses in Mining," AIME, *Trans.*, XXVIII (1898), 747.

33 "Provision for the Health and Comfort of Miners—Miners' Homes," AIME, *Trans.*, III (1874-75), 221, 225-27.

34 "The Hygiene of Mines," AIME, *Trans.*, VIII (1879-80), 97-120.

paternalism. The problem was deep-rooted and related to the fact that opportunities for advancement for the worker and the idea of the dignity of labor were fictions. Bayles's suggestion that the engineer arbitrate labor disputes was met with little enthusiasm in the AIME.[35]

In some contrast with the ASME and in deep contrast with the ASCE, the mining engineers' organization exhibited almost no interest in the questions of professionalization or professional status at all. This is not surprising in an occupational group which was more heavily entrepreneurial than mechanical engineering and in which recognition appears to have rested almost entirely upon entrepreneurial status. The low entrance requirements for membership further cast a shadow on the possibility of the institute's acting as a professional nucleus. Before 1908 the only mention in the institute's *Transactions* of the question of ethics was by James C. Bayles, who hardly qualifies as a mining engineer.[36] By 1908 such figures as John Hays Hammond were addressing themselves to the problem, but Hammond was himself an "engineer" before he was a mining engineer and was a leader in movements which cut across engineering lines.[37]

Mining engineering remained throughout this period a quasi-profession, a pariah to the civil engineers for starting the split into branches, a back-closet skeleton to the mechanical engineers, which they did not wish to repudiate, but which they clearly wished to rise above, and by its inaction and lack of professional activity a weight around the neck of those engineers who wished to raise the whole class of engineering occupations to the professional level.

Electrical engineering began as a division of mechanical engineering and was the first major component of that group to split off and form its own technical association. The ASME group was only a core membership, and within a short time after the formation of the American Institute of Electrical Engi-

35 J. C. Bayles, "The Engineer and the Wage-Earner," AIME, *Trans.*, XIV (1885–86), 327–36.

36 "Professional Ethics," AIME, *Trans.*, pp. 609–17.

37 John Hays Hammond, "Professional Ethics," AIME, *Trans.*, XXXIX (1908), 620–27.

View of the drafting room, Niles Tool Works, Hamilton, Ohio, 1891 (*Catalogue of the Niles Tool Works*).

neers in 1884 it had gained many members who had never been part of any previous organization. Electrical differed from mechanical engineering in that it had grown up as an occupation in the ten years preceding the founding of the AIEE. The occupational group was (compared to the other three major engineering divisions) born yesterday and had no long-standing tradition, no professional culture, and no antecedents in the ancient world. As a science it was in large part based on mechanical engineering. Within college curricula it grew up usually as an adjunct or addition to the program in mechanical engineering.

This fact led the mechanical engineers to regard electrical engineers as upstarts, but neither they nor the mining engineers were in any position to throw stones, having themselves broken engineering into parts. The new AIEE, moreover, made a strong case for the fact that it was a "sister science to civil engineering"[38] and was independent and distinct as a science and a profession. One electrical engineer addressed the AIEE, saying "the mechanical engineers claimed . . . electrical work as a branch of their profession. But the very existence of this body today evinces the uselessness of such a claim."[39] *Electrical World* asserted in 1892 that "an attempt to be both a mechanical and an electrical engineer cannot but result in failure. The fields are too large for any man to attempt to master both."[40]

Counter to such claims were articles like that in *Engineer* which declared electrical engineering not a distinct branch of the profession because such a large proportion of the work of the electrical engineer was almost entirely mechanical. The journal felt that, as a result, electrical would soon (1898) be reabsorbed into mechanical engineering.[41] Despite this kind of by-play, the exodus of electrical engineers in 1884 was not taken as a serious blow by mechanical engineers, as that of the mining engineers had been taken earlier by the civil engineers. Cooperation between the ASME and the AIEE was frequent and friendly, and no conflict seems to have existed.

---

38 Francis B. Crocker, discussion of Dugald C. Jackson, "The Technical Education of the Electrical Engineer," AIEE, *Trans.*, IX (1892), 487.
39 R. B. Owens, "Electric-Technical Education," AIEE, *Trans.*, IX (1892), 462.
40 Quoted by Arthur T. Woods, Letter to the Editor, *Amer. Mach.*, July 21, 1892, p. 7.
41 "Electrical and Mechanical Engineering," *Engineer* (N.Y.), May 2, 1898, p. 109.

Electrical engineers differed in the political organization of their technical association from the mechanical engineers in several particulars. Nominations were open to the floor and were not managed. Anyone could be elected to office, and younger men could reach high office much faster than in the other societies. For example, in 1888 Leo Daft and Charles Van Depoele were full members of the AIEE; they were the proponents of the two leading systems of electric traction then being experimented with widely in American cities. Frank J. Sprague, by contrast, was a practically unknown associate member. In the fall of 1888 Sprague introduced in Richmond, Virginia, the first technically and commercially successful system of electric traction mass transit, which, within the next four years, virtually swept the country. In 1892 Sprague was elected president of the AIEE. Such a skyrocketing into the presidency is without precedent in any of the three other major engineering societies.

A distinct lack of conservatism and concern for order was evident in the meetings of the institute. Proceedings were disorganized, lacked discipline, and discussions frequently got out of hand, leading to mass confusion. Perhaps some of this can be attributed to the fact that the members of the AIEE were somewhat younger than the members of the other three founder societies, but much of the cause can be laid to the fact that the electrical industry was in such a state of relative confusion and flux itself, with men and fortunes made and lost overnight, that the technical association representing its engineers could not help reflecting this. By the mid-nineties the AIEE began to become better organized, probably in part because of the addition of such outstanding electrical scientists as Michael Pupin and Charles P. Steinmetz, who lent character and dignity to the proceedings. Simultaneously, wide-open nominations were altered to include a printed ballot for "guidance of the members."

The absence of any shop, field, or mine culture in electrical engineering freed that field of the often bitter controversy between shop and school, theory and practice. Not that examples of antischool feeling were lacking, but the AIEE and its members were less plagued by this problem than were the other technical associations, with the possible exception of civil engineers.

Perhaps the absence of shop-and-school conflict was a result of the lack of what could be called a shop culture, coupled with the fact that electrical engineers, coming on the scene when they did in the 1880's and 1890's, tended to be technical school graduates and to take different kinds of roles in industry than did other types of engineers. As we have mentioned, the average age of members was lower than in other societies, with 34 per cent of the members in 1902 being 25 to 30 years old. An astonishing percentage of the AIEE members were salaried engineers in large electrical manufacturing and power-generating companies, and very, very few were entrepreneurs or even executives. Fifty-five per cent of AIEE members in 1903 were electrical engineers in manufacturing or operating companies, only 16 per cent were managers or superintendents or held other executive positions; the rest were scattered as consultants, teachers, draftsmen, and so on.[42] Not only did most electrical engineers work for companies as engineers and not as entrepreneurs, but by 1907 most of them worked for just two huge companies, General Electric and Westinghouse. *American Machinist* suggested that by that date these two companies so controlled the practice of electrical engineering that "almost insuperable barriers against new men with new ideas" were raised "and the practice became 'routine and stodgy.' " The electrical industry was remarkable for the speed with which it passed through various stages from entrepreneurial competition to consolidated bureaucracy, and this rapid change can certainly be blamed for some of the instability of the AIEE.[43]

The high percentage of technical degree-holders among the membership of the AIEE does not mean that technical graduates were necessarily leaders in the profession or in society in general. In a self-examination in 1910, the AIEE discovered that

---

42 Breakdown of AIEE by age and occupation in Charles F. Scott, "President's Address, June 29, 1903," AIEE, *Trans.*, XXII (1903), 6–7.

43 "Commercialized Engineering," *Amer. Mach.*, January 24, 1907, p. 125. Harold C. Passer's *The Electrical Manufacturers, 1875–1900* (Cambridge, Mass.: Harvard University Press, 1953) is a detailed analysis of the structure of the industry.

although 4.6 per cent of the population was engaged in the electrical industry, only 0.8 per cent were in *Who's Who* for 1906–7. This discovery led to the further discovery that 80.6 per cent of the membership held college degrees and 90 per cent had had some higher education. The former group, however, contained only 56 per cent of the AIEE members in *Who's Who*, while the latter group contained 70 per cent. The obvious inferences were not made and the rest of the relevant figures were not worked out, but the study indicates that the college degree-holders did not tend to be entrepreneurs. Early in its history as a professional association the AIEE was faced with the problem that plagued the other societies somewhat, the mechanical engineers' organization particularly—the danger that instead of a gentlemen's club for engineer-entrepreneurs the professional association might become a mass membership trade-union for young technical school graduates caught in white-collar jobs in modern bureaucratic corporations.[44]

Professional status for the organization was nearly established in 1885 to 1887, when the young AIEE began having financial difficulty and some members proposed creation of a large new class called junior associates, which would pay half dues, be nonvoting, and could include electricians and other practical men. A few members spoke out against such a tactic since they felt it would lower the dignity of the institute before the world. As one expressed it, "Don't let us confess that we are going out into the world to gather a few young men to get their three dollars a year." There seems to have been serious doubt that the organization could survive the financial crisis, until someone at that 1887 meeting began a series of spontaneous donations to the building fund for the new house and tens of thousands of dollars were pledged in a matter of minutes.[45]

Although it did not appear until about 1905, concern for professional ethics in the AIEE was considerable after that date. Charles P. Steinmetz pointed to the fact that "a very large per-

[44] Tabulation of AIEE members in *Who's Who in America* analyzed by age and education in Samuel Sheldon, "Education for Leadership in Electrical Engineering," AIEE, *Trans.*, XXIX (1910), 651–54.

[45] "AIEE Annual Meeting," AIEE, *Trans.*, II (1885), 1–17; "Business Proceedings of the American Institute of Electrical Engineers," discussion by Mr. Shelbourne, AIEE, *Trans.*, IV (1887), 26.

centage of the prominent electrical engineers are more or less closely associated with large manufacturing or large operating companies." This fact made a code of ethics very important, Steinmetz felt.[46] Steinmetz and other electrical engineers believed that their professional status was threatened by the situation. The appointment of a committee on the subject in 1906 led to extensive discussions later that year and the next about the whole problem of adopting a code of ethics for electrical engineering.[47]

The electrical engineers were, paradoxically, the most active group in pushing for engineering unity, not the inclusive unity of the type the civil engineers wished to create, but a federated unity with all branches of engineering in equal partnership. The last of three splinters wished to reconstitute the tree. By the early twentieth century the AIEE was going out of its way to invite other types of engineers to its meetings and was trying to make the AIEE a common meeting ground for them all.[48] Charles F. Scott was the major figure in the movement for engineering unity, and it was after hearing an impassioned plea for that cause at the AIEE Library Dinner in 1903 that Andrew Carnegie gave the AIEE the chance to play a major role in a project leading toward that end.

Concern for engineering unity began to be shown by engineers of all types shortly after the fact of disunity had made itself clear. The decade of the 1880's saw campaigns mounted to reconstitute the divided profession and to establish some kind of American engineering organization which would include the best men from each of the four founder societies and specialties and which would give engineering a single, unified voice in the councils of industry and the nation. By 1884 three organizations based in Chicago were working toward this end, the journal *American Engineer*, the Association of Engineering Societies, and the Western Society of Engineers.

Specifically, *American Engineer* called for a great American

---

[46] "Discussion on 'Engineering Honor' at Milwaukee, Wis., May 30, 1906," AIEE, *Trans.*, XXV (1906), 266.

[47] Report of Committee on "Proposed Code of Ethics," AIEE, *Trans.*, XXVI (1907), 1422–25.

[48] Calvin W. Rice, "Institute Branch Meetings, Their Organization, Development and Influence," AIEE, *Trans.*, XXII (1903), 64–66.

association of engineers, "specialized in sections, and localized in chapters; representative, dignified, a powerful social and political unit, moulding public thought, projecting great enterprises, and leaving its impress on legislation."[49] Local and regional societies, like the Chicago-based Western Society of Engineers, which published their transactions in the Association of Engineering Societies *Journal*, echoed this cry, and President Benezette Williams remarked on his inauguration in January, 1885, that the infant Association of Engineering Societies was not strong enough to accomplish the task of unifying the profession. The American Society of Civil Engineers was perhaps equal to the task, according to Williams, but it seemed, on the basis of past performance, unlikely to do it.[50] Though he could not foresee it, by the time the ASCE chose to play this role, no one else was willing for them to have it. This whole movement, connected as it was to the three Chicago-based operations, was likely a bid to steal the thunder of the New York-based national engineering societies. As *American Engineer* expressed it in 1885, "the plan would have a decentralizing effect and would therefore incur the opposition of those agglutinated about the hub," meaning, of course, New York City.[51]

With a few exceptions, before 1893 the major support for the idea came from the local engineering societies in places like Chicago, St. Louis, and Cleveland. Presidents of the Engineer's Club of St. Louis and of the Civil Engineer's Club of Cleveland called for the formation of a truly national engineering body, looked to the strengthening of the Association of Engineering Societies as one answer (a federal body, with representatives from only the Midwest), and stressed the fact that any such body must not be unitary with one central headquarters.[52] In 1886, William Kent proposed before the American Association for the Advancement of Science that a great acad-

---

49 "Engineering Societies," *Amer. Engr.* (Chicago), November 14, 1884, pp. 195–96.

50 "Inaugural Address of President Benezette Williams to Western Society of Engineers, January 20, 1885," AES, *Jour.*, IV (1884–85), 130–33.

51 "Engineering Societies," *Amer. Engr.* (Chicago), January 16, 1885, p. 25.

52 William H. Searles, "The Outlook for Local and General Engineering Societies," given to Civil Engineer's Club of Cleveland, AES, *Jour.*, X (1891), 196; William B. Potter, "Address on Retiring from the Presidency of the Club," given to Engineer's Club of St. Louis, AES, *Jour.*, VII (1888), 25–28.

emy of engineering be created, numbering in membership "not over 300 or 400, but powerful in its influence for good, whose members comprise army and navy engineers, civil, mining, metallurgical, mechanical, electrical and sanitary engineers, and all are distinguished men in their separate branches, having *done* something creditable in the engineering line as a prerequisite to membership. It is an aristocracy based upon intellect and achievement."[53]

The idea was not really discussed in the ASME until Robert Thurston brought it up in 1889 at the New York meeting. He recalled his memories of the great departed engineer, Alexander Lyman Holley, who, Thurston claimed, believed in the idea of an academy of engineering, having as members the best of every specialty. Paradoxically, Holley was the man who did as much as Thurston to create engineering disunity, by aiding the foundering American Institute of Mining Engineers in 1875 and by founding the American Society of Mechanical Engineers in 1880. Thurston's suggestion brought some approval but much criticism. J. F. Holloway, pump designer and expert with H. R. Worthington's firm, did not want to see the ASME "sunk into another society." W. S. Rogers expressed the same conviction,[54] and Daniel Ashworth thought the idea un-American and "undemocratic . . . and . . . believe[d] it would create a spirit of caste—a special class."[55]

As early as 1890, engineers of all kinds who fancied the creation of a master engineering society found a practical vent for their enthusiasm in the agitation that developed for a union engineering building in New York City. Such a building would be large enough to house the ASCE, AIME, ASME, AIEE, and several other organizations, the most frequently mentioned being the Engineer's Club of New York. It was one of the issues discussed by a joint committee appointed by the four founder societies in 1889 in response to Thurston's appeal of that year.

53 "Proposal for an American Academy of Engineering," *Amer. Engr.* (Chicago), October 6, 1886, p. 132.

54 Thurston, Holloway, and Rogers, discussion, "Proceedings of the New York Meeting," ASME, *Trans.*, XI (1889–90), 33–36, 41. Oberlin Smith reported to the ASME Council in 1890 that the civil engineers were opposed to the idea of an institution of engineers. ASME Council Minutes, July 9, 1890.

55 "Proceedings of the Cincinnati Meeting," ASME, *Trans.*, XI (1889–90), 601.

Although the plan of this committee for steps toward unification brought an outright refusal by the civil and mining engineers, the mechanical and electrical engineers seemed to be in agreement on the question of the desirability of unification and the construction of a union headquarters.[56]

The Chicago World's Fair, which many engineers attended, led to considerable discussion of the idea, and it was picked up and exploited by the technical periodicals in the decade that followed. By 1901 *American Machinist* was lyrically writing, "It stirs one's blood to think of the building which they [united engineers] could maintain and of the influence which it would exert on the engineering profession and on the estimation of the profession by the general public."[57] Mechanical engineers, with some exceptions, generally supported the idea, but the electrical engineers were the project's biggest boosters. Charles F. Scott, a power in the AIEE, talked about almost nothing else but engineering unity and the advantages of a union building. It was after one of his speeches on the subject that Andrew Carnegie, a guest of the AIEE for their Library Dinner on January 10, 1903, offered one million dollars to the four founder societies to construct a union engineering building and summoned Scott and Calvin W. Rice to his office to discuss the gift.[58]

For the next six months plans for the building were discussed in the engineering societies and in the technical press. Electrical engineers were quite enthusiastic, with mechanical engineers running a close second. After some misgivings, related apparently to their desire to avoid a large-scale financial commitment and headquarters operation, the mining engineers came in. Only the civil engineers were left, and speculation ran that they would reject the plan because of their investment of $250,000 in a house several years earlier.[59] They had not responded to a call

---

[56] "Proceedings of the Richmond Meeting," ASME, *Trans.*, XII (1890–91), 6–7. As early as 1888 the AIEE was negotiating with the ASME concerning the possibilities of a joint building. ASME Council Minutes, April 6, 1888.

[57] "A Union Engineering Society House," *Amer. Mach.*, August 22, 1901, p. 935.

[58] AIEE, *Trans.*, XXI (1903), 495; "Library Dinner of the AIEE," AIEE, *Trans.*, XXI (1903), 98–99.

[59] "The Union Engineering Building," *Amer. Mach.*, May 14, 1903, p. 708.

by the *American Machinist* in 1901 for joint ownership of that house.[60] It does not appear to have been the investment in their new house that troubled the civil engineers about accepting the Carnegie offer, but rather the fear of contamination by contact with a lower order of professional associations, and most particularly with the mining engineers' organization. *Engineering News*, the civil engineers' leading periodical, carried articles which fanned the flames of almost buried rivalries, and J. J. R. Croes wrote in that journal: "The only reason for the existence of the other societies is that there have been and always will be a good many men who feel the need of an association; but are either unable or unwilling to comply with the requirements of the American Society of Civil Engineers."[61] To this renewal of the idea of civil engineering's inclusiveness, *American Machinist* hotly replied that the civil engineers had an exaggerated idea of their own importance. The journal went on: "This building is certain to become the recognized engineering center of the country, and if the Civil Engineers remain without the circle they must be content to see their headquarters, and with them their society, relegated to a secondary position, known and recognized by engineers, but unknown to anyone else."[62]

The civil engineers' organization sent out a letter ballot on the question and rejected the ASCE's participation in the union building by a vote of 1,139 to 662. Carnegie did not withdraw his gift, and the project went ahead with the three other societies participating. The building was dedicated in 1907 and did become the center of engineering as predicted. The event marked the emergence of the ASME as a distinct challenger to the leadership of the ASCE, and this was recognized when the latter organization finally did come into the union building in 1918, eleven years after the dedication. At the time of the

60 "An Opportunity for Co-operation Between Engineering Societies," *Amer. Mach.*, January 24, 1901, p. 78. They were perhaps justified in taking a negative attitude toward the union building idea. In 1887 the ASCE had suggested to the ASME that they jointly purchase a large mansion then on the market for a library and society building to serve both organizations. The ASME Council declined on account of the location of the property. ASME Council Minutes, April 27, 1887.

61 Quoted in "The Union Building for Engineers—A Business Proposition," *Amer. Mach.*, June 11, 1903, pp. 850–51.

62 "The Union Engineering Building and the Civil Engineers," *Amer. Mach.*, p. 850.

rejection, however, tempers ran hot in the technical journals. *American Machinist* reported the vote, suggesting:

Here was a plain business proposition which could be considered properly upon strictly business principles only. It was not professional, it was not social, it contemplated no amalgamation of the societies nor any joint action as societies either professionally or otherwise.

To be consistent the civil engineers should now vote for the erection and constant maintenance in New York of a hotel large enough to accommodate their entire membership in attendance at any convention to be held there and not to be used by any other persons at any other time, especially by persons who may be members of any other engineering organizations or otherwise masquerading as engineers.[63]

At the first statement of this lampoon one might easily imagine the civil engineers shouting in unison that professionals might have good reasons for rejecting a "business" proposition.

Despite the occasional unwillingness of the civil engineers to cooperate, the founder societies greatly increased their areas of united action in the first decade of the twentieth century. Much of the important action formulated and carried out by engineers in this period and after was joint action. A good example was the joint response to President Theodore Roosevelt's call for action to promote efficient use of the nation's resources in 1908.[64] Although able to unite on such outside issues as conservation, the founder societies were never able to come to any kind of agreement on the creation of a general profession of engineering, with minimum standards of education and experience required for the right to practice. "Engineer," with a capital "E," remained an abstraction trotted out for yearly speeches. Engineering unity was a myth, not a reality.

[63] "The Carnegie Union Engineering Building," *Amer. Mach.*, March 10, 1904, pp. 334–35.
[64] "Meetings, January-June," ASME, *Trans.*, XXXI (1909), 7.

George W. Light correctly pegged the nature of the art and science of mechanical engineering in 1835 when he referred to it as among the "business professions." Light went on to point out that mechanical science was narrowing the gap between art and science and that the progress of steam power was creating the impetus for this movement.[1] Over the next seventy-five years mechanical engineering became infinitely more scientific in nature, developed certain professional attributes, formed national and local professional associations, developed professional schools, but in all of this never lost sight of the fact that it was first of all a business.

The school and shop cultures that clashed on the question of which institution was best to educate and socialize the mechanical engineer could both agree that he was no engineer at all if he did not have the terms of profit and loss constantly in view. Henry R. Towne told the students at Purdue in 1905 that "the dollar is the final term in every engineering equation."[2] Coleman Sellers, educator, scientist, and shop culture elitist, told the ASME as president in 1887 that "we must measure all things by the test, *will it pay?*"[3] Some professionally oriented engineers might gasp at the remarks of the president of the Stevens Alumni Association in 1896: "It must be remembered that the financial side of engineering is always the most important, and that the sooner the young engineer recedes from the idea that, simply because he is a professional man, his position is paramount, the better it will be for him. He must always be subservient to those who represent the money invested in the enterprise."[4] Few, however, would have had the courage to challenge such a doctrine in the forum of the ASME. Many engineers believed that such a situation was far from limiting,

---

[1] "A Plea for the Laboring Classes," *New England Magazine*, reprinted in *Boston Mechanic*, IV (December, 1835), 239.

[2] "Industrial Engineering," *Amer. Mach.*, July 20, 1905, p. 100.

[3] ASME, *Trans.*, VIII (1886–87), 695.

[4] "The Engineer as a Business Man," *Amer. Mach.*, August 6, 1896, p. 752.

but instead gave guidelines that kept them in contact with reality. The hoped-for end product of this union of scientific profession and profit logic was the engineer-entrepreneur.

There was, however, nothing in the image of the engineer-entrepreneur that smacked of money grubbing. It was not necessary for the engineer to amass a large fortune in order to be acceptable in the elite—it was necessary that what he did was profitable, however small-scale or experimental it might be. Although there are many examples of the engineer-entrepreneur among the membership of the ASME, one name was frequently mentioned as representing the ideal (not typical) representation of the type. Alexander Lyman Holley was perhaps best known (among engineers) for his experimental plant in Troy, New York, which made Bessemer steel. He was not the first man to make Bessemer steel in this country, nor was he the man who made the most money at it; he was instead the first man who made it at a profit. This was no accident. Holley had this purpose firmly in mind when he began the task; as William Metcalf eulogized him after his premature death in 1882: "He seized upon the great Bessemer process when capitalists were afraid, when old practices and ignorant prejudice derided it, and when a slumbering world had no idea of its own needs."[5]

Holley himself was one of the earliest to formulate the engineer-entrepreneur as an ideal. Educators also gave lip service to it, even though they did not often promote it directly. Professor Thomas Egleston of Columbia University defined engineering as "the science of making money for capital. Its two essentials are *honesty of purpose,* and great care to secure *accuracy of results.*"[6] Dugald C. Jackson, engineering educator, told graduating engineers at the University of Colorado that the professional engineer must be "a man of science, a man of the world, a man of business."[7] Almost unanimously engineers

5 "A Tribute to Alexander Lyman Holley," ASME, *Trans.,* III (1882), 52.

Fritz Redlich was one of the first to treat the engineer-entrepreneur in his *History of American Business Leaders* (2 vols.; Ann Arbor, Mich.: Edwards Bros., 1940). John B. Rae has investigated this figure in "The Engineer as Business Man in American Industry," *Explorations in Entrepreneurial History,* VII (December, 1954), 94–104, and "The Engineer-Entrepreneur in the American Automobile Industry," *Explorations in Entrepreneurial History,* VIII (October, 1955), 1–11.

6 *Amer. Mach.,* May 28, 1881, p. 8.

7 "The Potency of Engineering Schools and Their Imperfections," *Wisc. Engr.,* VII (May, 1903), 154.

of all types agreed that whatever else engineering was, it was first of all a business.

Robert H. Thurston, who deftly walked the tight wire between shop and school cultures, while promoting whenever possible the latter, pointed this out clearly during his year as first president of the young ASME. "I presume no association contains so large a proportion of men who are tied to business as the members of this Society," he told the members in 1881. The society member could deal with "great social problems" in his role "as an employer and a director of labor and capital." The new society would, Thurston believed, enable "business men," who hitherto had had little influence on national legislation, to act in an organized fashion.[8]

Mechanical engineers were, however, a particular kind of businessman, and as such saw their world from a different perspective than did other types. One broad area in which few mechanical engineers were employed was primary extraction of raw materials. This meant that they did not as a rule come into contact with large groups of exploited or otherwise downtrodden wage-earners, such as miners, or with large-scale violence requiring organized and efficient repression by entrepreneurs and superintendents. The mining engineers, who did have such contact, were, for example, more alarmed about the dangers of socialism, trade-unions, and any attempts of the workers to organize than were the mechanical engineers. In the machine shop, often the worst that one had to cope with was an occasional tramp machinist stirring up the men.

Secondary manufacturing and processing industry was the broad, general area in which the mechanical engineer was employed. Within this area, however, he was much more likely to be found in those industries which were industry oriented than in those which were consumer oriented. The consumer-oriented industry is often much more unstable (at least in its early stages) than is that which sells its products to industry. Consumer industry must—at least before market research surveys—unscientifically attempt to gauge the whims and fancies

---

8 "Proceedings of the Hartford Meeting," ASME, *Trans.*, II (1881), 3–4; Robert H. Thurston, "The Mechanical Engineer—His Work and His Policy," p. 76; "President's Inaugural Address," ASME, *Trans.*, I (1880), 3–4.

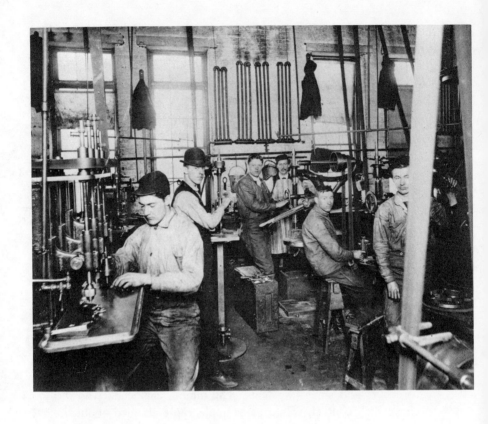

The shop at the C. H. Mergenthaler and Co., Engineers and
Machinists, Baltimore, Md., 1894.

of the public. Competition, before the emergence of trusts, pools, and other forms of industry-sponsored control, is generally brutal and relentless. Frequently, market considerations take precedence over engineering considerations. Engineering work is often a one-time affair, concerned with building a plant for the manufacture of a product and requiring large-scale, long-term investment that precludes frequent changes of machinery, even if better machinery is later invented and perfected. A good example is cotton cloth manufacture and its sister industry, the garment-making industry. Such industries tended to have low levels of innovation (in the area of machinery for production) and did not, in the period under discussion, generally offer employment to mechanical engineers. By the 1890's, however, many consumer industries were hiring technical graduates as "mechanical engineers" for their plants.

The members of the elite who led the ASME worked in the industry-oriented machine shops, engine shops, pump and valve works, foundries, instrument and gauge plants. This is really quite a restricted segment of industry. It is small in dollar volume, numbers of employees, total capitalization, and other common measures of importance used by the economist. It makes the tools, fittings, gauges, and other basic equipment which can be used to produce articles for sale in the consumer market and thus represents a theoretical and frequently an actual bottleneck in the flow and process of industrial growth. Entry into the business requires only a moderate amount of capital (a fact that allowed simple partnerships to dominate the form of business organization into the twentieth century), but requirements of know-how are very high, restricting entry to those individuals who possess technical skill, some capital, and a generous helping of business acumen. Thus the engineer-entrepreneur is almost a *sine qua non* of these industries.

In the late nineteenth century, at least, this industrial cluster was one in which *who* one knew counted for nearly as much as *what* one knew. Contracts for basic machinery were let on a personal level, and the shop entrepreneur frequently knew the directors of the manufacturing industry and was able to gear his innovation in basic tools, engines, and so on, to the specific needs of that customer. Competition was restrained and gentlemanly for the most part and was kept that way by the market conditions. The level of innovation was high, and machine-tool shops often kept one jump ahead of the manufacturers' needs. Under these protected market conditions innovation was less risky. Risk was kept down particularly by the fact that many of the shop owners were actually on the boards of directors of the companies to which they sold. A good example of this phenomenon was William Sellers of Philadelphia, who sat on the board of the Pennsylvania Railroad, one of his biggest customers for machine tools.[9]

Thus mechanical engineering was entrepreneurial, but within distinct limits. Though entry was easy from a financial standpoint, technical requirements and the need for personal connections played an inordinately large role in structuring and controlling the industries in which the mechanical engineer found employment. Mechanical engineering produced many millionaires, no doubt, but very few multi-millionaires. The industrial position of its practitioners was secure and respectable, but certainly not all-powerful.

For obvious reasons mechanical engineers did not find employment in tertiary or commercial and service industries. If a mechanical engineer had become a department store king, his fellow engineers would certainly not have continued to admit that he was a member of the club. Mechanical engineering was a business, but a very particular kind of business. By the 1890's, however, mechanical engineers were entering other kinds of business and industry in large numbers. The shop culture complex could not absorb the many technical school graduates who appeared on the employment market in the late nineteenth and early twentieth centuries. Even many of the small machine shops began to expand in size and number of employees, to incorporate, and ultimately to bureaucratize. Mechanical engineering remained a business profession, but became one of somewhat broader scope and experience.

Two questions arise from this discussion: Did the shop culture ideology insist that it was necessary for the engineer to be also an entrepreneur, or at least to have entrepreneurial sense?

9 Strassman, *Risk and Technological Innovation*, pp. 116–57.

Was entrepreneurial sense necessary for the successful practice of mechanical engineering? The answers are that both in ideology and practice it was almost always necessary for an engineer to leave the engineering of materials and enter the engineering of men in order to become very successful financially and socially in the terms prescribed by the society in which he lived. In a sense it was necessary for the engineer to leave engineering proper for management and executive positions in order to expect the kind of deference which the more sanguine hoped he might achieve as a "pure" engineer. The mobility route to elite positions in American society was to leave pure engineering. This was a tendency that increased, if anything, with the advent of the technical graduate in large numbers.

American Machinist printed a letter by an instructor, which posed the problem in these terms: "A man who would be an engineer [of materials] must be content to be a plodder so far as wealth is concerned. The greatest wealth is gained by those who can engineer men."[10] John F. Hayford told the graduating class of the Clarkson Memorial School of Technology in 1907 that "to attain the highest success as an engineer you must not be the type of man who knows how to do things excellently but cannot tell others how to do them. . . . Instead of devoting your energy simply to increasing your own output by fifty or one hundred per cent, it is far better . . . to increase the output of each of one hundred men by ten per cent. The world recognizes this by awarding the prizes to the administrators."[11] A civil engineer, addressing the profession as a whole, spoke in 1892 with some bitterness about the situation. He was forced to the conclusions that the "Engineer, to amass a competency, must . . . abandon his profession, as thousands have done already, becoming in time Railway Presidents, Lawyers, Statesmen, Experts in Insurance, Managers or Men of Affairs generally." Discussion followed this talk at the Civil Engineer's Club of Cleveland, and the consensus of the civil, mechanical, and electrical engineers present was that the engineer must become a businessman to be counted a success. Most agreed also that this was desirable, but one engineer, Ludwig Herman,

10 "Technical Education," *Amer. Mach.*, March 2, 1905, p. 300.
11 "Study Men," *Addresses*, ed. Waddell and Harrington, p. 348.

criticized a civilization in which all an engineer needed to get ahead was "a little knowledge with a great deal of brass."[12]

The large corporations which were hiring the young technical graduates right out of school made this a part of their sales pitch. By 1925 Westinghouse claimed that since 1895 they had been recruiting "executives and technical experts from among those who show capacity and who enter the organization directly from engineering schools." According to the company, this included most of the "vice-presidents, its departmental managers, district sales managers, works managers, service managers." The main attraction in working for the company was thus presented as not the fact that one could work as an engineer, but that engineering offered a sure route into desirable executive positions.[13]

Engineering educators frequently pointed out that they were training men not merely to be engineers, but to be sharp businessmen as well. Complaints from shop culture often scored the lack of business training in the technical schools' curriculum. Both agreed, however, that it was necessary for real success. Shop culture's complaint about the schools was expressed clearly in 1905 by a telephone engineer, Angus S. Hibbard: "A young engineer entering upon his work is so full of the technical side of the business that he finds it difficult to assimilate with the corporate body of engineers in which he finds himself." Such an engineer failed to acquire executive ability, Hibbard suggested, and "the most valuable development of the engineer is up through his engineering work into an executive position." The fault lay with the schools, which simply were not training good leaders and administrators.[14] Bell Crank, an anonymous correspondent of the *American Machinist*, felt that a common fault of technical graduates was their lack of "business knowledge and qualifications."[15]

A frequent complaint was that the technical school did not

---

[12] Leon Gobeille, "The Financial Status of the Engineer," and discussion, AES, *Jour.*, XI (1892), 495.

[13] Westinghouse Electric Corporation, *The Engineering Graduate* (Pittsburgh: The Westinghouse Corporation, 1925), p. 6.

[14] *Wisc. Engr.*, IX (April, 1905), 178.

[15] "Development of a Mechanic," *Amer. Mach.*, March 10, 1892, p. 5.

teach the prospective engineer the difference between mechanical and financial efficiency. A sample of this kind of problem, which faced mechanical engineers on a regular basis in their day-to-day relations with business and industry, is the following: Assume the engineer has a choice of two rates of burning coal to use in firing the boiler of a railroad locomotive. One rate is mechanically more efficient—that is, it gets more power per pound than another rate, thus is less expensive in fuel per mile. Let us say that burning the coal at a higher rate will increase the steam power generated by only one-third for half again as much coal consumed. According to H. Wade Hibbard, the green young mechanical engineer would advise use of the lower, more efficient burning rate, forgetting or never knowing that a one-third increase in speed and hauling power for the locomotive represents a tremendous saving since "the extra coal is almost the only cost of hauling the extra cars because no increase is made in the interest on locomotive, track and bonds, in the wages of that engine and train crew, or of the track and signal men, in salaries of officials, in taxes or insurance."[16]

E. D. Leavitt told the ASME that in the situation where he was given complete freedom in selecting an engine for a steam installation his "reputation as an engineer would not allow [him] to put in a more costly [and better from an engineering standpoint] machine when the advantages would not save a handsome interest on the extra cost."[17] True mechanical efficiency, abstractly conceived, might include the idea of making use of every waste and by-product of manufacture. But, as *Industrial Monthly* pointed out, "no man can afford to spend time and money in the effort to find a market for these [waste materials and by-products] unless they are produced in large quantity and are well known in commerce."[18] *American Engineer* derided the type of theorist who "seeks to attain at a monetary sacrifice a mathematical perfection in machines and appliances which is not required by the specific demands that the

16 "Railway Mechanical Engineering," p. 106.
17 "Discussion," ASME, *Trans.*, II (1881), 205.
18 "Waste Materials and Their Utilization," *Industrial Monthly*, I (February, 1872), 55.

mechanical device is to meet." No experienced engineer would do such a thing and tell about it. Young engineers were apparently the most prone to this type of fault, particularly those with only a technical education.[19]

A few engineers, however, did see possible and actual conflict between the engineer as engineer and as businessman. Even such a proponent of shop culture as Alex Holley recommended the class of associate member for businessmen in the ASME, to maintain the "professional standard."[20] *American Engineer* suggested in 1882 that the object of technical schools was the teaching of a profession "from the aspect of a science, and not only from the aspect of the art of money making."[21] Some, like C. L. Redfield, a frequent contributor to the *American Machinist*, raised serious questions about the "relationship which the mechanical engineer, as a professional, bears to the business public."[22] Redfield attacked the growing field of engineering sales, suggesting that just because some manufacturers' agents were mechanical engineers did not give all of them the right to so call themselves. Nor was the idea that one had to be a manufacturer to be an engineer a good one for the profession, Redfield thought.[23] Engineering educator John Lyle Harrington believed that "it is rare indeed that the engineer is free to act according to his best judgment, no matter what his position. If he be in the employ of a manufacturer or a contractor, competition forces him to adopt many methods which fall short of the best."[24]

Only one full-scale exploration of the problem of the relationship between engineer and businessman was presented to the ASME during these years. William H. Bryan spoke in 1898 on "The Relations between the Purchaser, the Engineer, and the Manufacturer," but he dealt only with the problem as it affected

---

[19] "Commercial Engineering," *Amer. Engr.* (Chicago), August 20, 1885, p. 71. Colvin, *Sixty Years*, p. 100, and Hitchcock, *My 50 Years*, pp. 66–68, have examples of the way experience was used to temper logic and science.
[20] "The Field of Mechanical Engineering," ASME, *Trans.*, I (1880), 5.
[21] "Technical Education," *Amer. Engr.* (Chicago), December 8, 1882, p. 263.
[22] "What Is a Mechanical Engineer?" p. 5.
[23] *Ibid.*, p. 6.
[24] Editorial comment on Waddell, "Address to the Graduating Class of the Rose Polytechnic Institute," p. 417.

consulting engineers.[25] In truth the elite mechanical engineers did not consider the relationship a problem. It existed and was necessary to the survival of mechanical engineering as an occupation or a profession. Besides, as shop writer F. F. Hemenway wrote in the *American Machinist*, if a mechanical engineer really wished to revolutionize industrial practice he could become an entrepreneur, a "high and mightly engine builder, who could [as could a William Sellers, for example] compel a demand for what they make. They educate ignorance up to demanding what is best."[26] Mechanical engineering was a business and those who would succeed in it were not allowed to forget that for long.

It was in their role as businessmen that mechanical engineers first became interested in the rational and scientific management of the workshop. Scientific management is a convenient but misleading term. It is used to refer to all aspects of the general movement to rationalize production and manufacturing in the era from 1890 to 1925, and it is descriptive of the general characteristics of the effort. Though the term will be used in this section, it is meant to refer to all efforts to rationalize industry and make it conform to the image of the machine. The formation of the ASME in 1880 presented a possible forum for discussion of the ways and means of so doing, and mechanical engineers were not slow to take advantage of it.[27]

American Machinist spelled out in 1882 its editor's ideas of the needs of mechanical engineering industry:

Above all what is needed is the incorporation of the *science* of machine shop economy into the system of machine shop instruction. Let it be understood that the training of an apprentice should be such as to fit him in the future to be . . . as strictly accountable for his part of machine shop economy as the manager. . . . There is a comparatively new field in the direction indicated, which, properly cultivated, will materially change the prospects of many machine enterprises.[28]

---

25 ASME, *Trans.*, XIX (1897–98), 686–99.

26 "Shop Esthetics," *Amer. Mach.*, June 12, 1880, p. 5.

27 For a discussion of the ASME's relation to scientific management, see American Society of Mechanical Engineers, *Fifty Years Progress in Management, 1910–1960* (New York: American Society of Mechanical Engineers, 1960).

28 "Machine Shop Economy," *Amer. Mach.*, July 22, 1882, p. 8.

Throughout the 1880's these ideas were followed up by engineer-managers and entrepreneurs in the form of systems of cost accounting, wage-payment plans involving premiums, and bonuses for increased production. Dozens of such plans found their way into the pages of the *American Machinist* in these years. Their objective was that the informal system of allowing the foreman personally to judge whether each man in a shop was doing his best be replaced with some systematic, impartial, and rational system of job or time cards. Workers could then be offered incentives or premiums for turning out more than a normal day's production. Obviously the problem was how to determine the true measure of an honest day's work and how to structure the premium plan so as to prevent cheating. These problems and others led some mechanical engineers to create a new field, the science of management, which quickly broadened its scope of problems and interest to the entire subject of industrial production in its every detail.[29]

In the early stages of scientific management, the wide membership of the ASME found much in it that was useful to them. In a very real sense the mechanical engineers' organization might be said to have spawned and nurtured the young profession until it was ready and strong enough to attempt the reform and rationalization of all American industry. Many of the innovators in this field were prominent members of the ASME, such as Henry R. Towne, who delivered before the group in May, 1886, a paper which called for the engineer to adopt a role as economist in the process of rationalizing the shop. Management techniques, Towne concluded, had no literature and no professional associations. Why should not the ASME form an "Economic Section" to deal with such problems, share knowledge, and develop standard procedures and forms.[30]

Not everyone agreed with this idea. As *American Machinist* noted, the paper was not, "strictly speaking, of an engineering character," and the editors judged the meeting a poor one. The

---

[29] For more information on scientific management movements, see Samuel Haber, *Efficiency and Uplift: Scientific Management in the Progressive Era, 1890–1920* (Chicago: University of Chicago Press, 1964); Horace Bookwalter Drury, *Scientific Management* (3rd ed.; New York: Columbia University Press, 1922).

[30] "The Engineer as an Economist," ASME, *Trans.*, VII (1885–86), 428–32.

journal noted that, in relation to Towne's suggestion of a section for economic or management engineering, "happily, we think, the members showed no inclination to move in the direction indicated."[31] Part of the reason for the lack of interest in the idea on the part of the members of the ASME can be attributed to their opposition in principle to sections of any kind. There was, however, a deeper conflict and challenge which the ideology and practice of scientific management offered to the shop culture ideal.

The details of the rival plans for the rational management of shops developed by Frederick Winslow Taylor, Henry R. Towne, Henry L. Gantt, Fred Halsey, and others are not relevant here except inasmuch as they reveal the relationship between the engineering and business communities and illustrate the conflict of values that the movement generated. As businessmen one might expect the mechanical engineering elite to have embraced scientific management with open arms and as engineers one would expect a similar reaction. That all did not react this way can be explained partly by the observation that by stressing piecework most of the management plans tied the machinist in the shop to one machine and placed a premium on his becoming extremely fast and proficient at turning out one particular kind of work. It attacked the notion that it was any longer possible for the versatile, creative mechanic to appear in the shop, work his way through its structure, acquire education, and become a mechanical engineer. It purported to replace the creative individual entrepreneur having intimate knowledge of his customers, his men, and his machines, ruling his shop with an iron yet personal hand, with an impersonal system, designed to transcend the competence of the individual foreman or entrepreneur. The management systems made it possible, ideally, for mediocre or average men following the system to administer a shop more successfully than the most gifted individual not using the system.

The kinds of objections raised to the various competing management plans are instructive. Frank H. Richards suggested that "just treatment, and nothing but just treatment, will beget

---

31 "The Mechanical Engineer at Chicago," *Amer. Mach.*, June 19, 1886, p. 8.

faithful service."[32] Fred Halsey, defending his plan of management before the ASME, retorted to such comments that it was simply not in human nature to be just and fair. John T. Hawkins challenged the Halsey plan on the grounds that his premium plan allowed only for increased pay for productive workers, but made no provision for savings for the shop directly or for the consumer in the form of lower prices for machinery and other engineering products.[33] Another tack was taken by *Engineer*, which pronounced the bonus system impracticable for use in power plants because of the small number of men involved and the close personal control that prevailed.[34] W. S. Rogers expressed a similar criticism in an ASME meeting that there were "many concerns running forty or fifty men that cannot afford to pay a $10,000 expert to run their business and have anything left."[35] *Engineer* noted that most shops were overmanaged and that they had never seen a man well treated who did not do a fair day's work.[36] *American Machinist* through the nineties took a strong stand against any system which relied on piecework, noting with some warmth: "It does hurt some managers if they see their men having a little easy time. They like to see their men sweltering in sweat, and with half their clothes torn off."[37]

Criticism of scientific management was obviously somewhat diffuse and lacked clear alternatives. Nevertheless, a feeling of despair does come through—something small, personal, and possessed of variety and opportunity was being threatened by something large, impersonal, and characterized by monotony and tyranny of a system. One thing on which even the critics of management engineering could agree with the defenders was

[32] "Human Nature in the Machine Shop," *Amer. Mach.*, August 2, 1894, p. 7.

[33] Comment on F. A. Halsey, "The Premium Plan of Paying for Labor," ASME, *Trans.*, XII (1890–91), 765, 774, 777.

[34] "Bonus System Impracticable in Power Plants," *Engineer* (N.Y.), January 15, 1902, p. 57.

[35] Comment on Frank Richards, "Gift Propositions for Paying Workers," ASME, *Trans.*, XXIV (1902–3), 275.

[36] "Shop Management," *Engineer* (N.Y.), October 13, 1894, p. 90.

[37] Pieceworker, Letter to the Editor, "Piecework," *Amer. Mach.*, August 9, 1894, p. 6. The resistence to the idea on the part of many businessmen-engineers was accurately pegged by Morris L. Cooke in a letter to Frederick W. Taylor in 1907. Cooke wrote that businessmen are in business not only to make money but to say how it is made and how business is conducted. Cooke to Taylor, October 7, 1907, in Taylor Collection.

that such systems might reduce the danger of trade-unionism among machinists and other workers. Robert Thurston noted that "our present common wage-system must inevitably be improved upon in the coming decade if we are to avoid very grave disturbances in our social system."[38] Colonel E. D. Meier spoke in the ASME in favor of bonus systems on the basis that "we cannot ignore the fact that labor unions tend to restrict production by making the pay of the poorest workman the same as that of the best, which kills individual ambition and energy. I think the bonus system a very practical way of meeting that."[39] Harrington Emerson, an exploiter rather than an innovator in scientific management methods, suggested to the ASME in 1904 that "I cannot but regard disputes over wages as the effect rather than the cause of unsatisfactory and unscientific relations between employer and employee."[40]

Nevertheless, it was still a long way from the rigid and inflexible rules and system which prevailed in the lock plant of Henry R. Towne to the genial suggestion of Oberlin Smith that what was necessary for better relations was for the boss to talk with and "take a glass of beer with" the trade-union representatives.[41] Basic internal conflicts in scientific management itself and in its adherents were not often recognized by the individuals involved. Frederick W. Taylor, after whom the movement is often called, expressed both admiration for and annoyance with the prevailing personal methods of shop administration. In his seminal paper, delivered before the ASME in 1895, "A Piece-Rate System, Being a Step Toward Partial Solution of the Labor Problem," Taylor criticized managers for believing in "men, not in methods," but later in the paper attacked the same managers for their lack of personal com-

[38] Comment on H. L. Gantt, "A Bonus System of Rewarding Labor," ASME, *Trans.*, XXIII (1901–2), 361.

[39] Comment on F. Richards, ASME, *Trans.*, XXIV (1902–3), 273.

[40] "A Rational Basis for Wages," ASME, *Trans.*, XXV (1903–4), 878. On the general problem of scientific management's relationship to labor, see Milton J. Nadworny, *Scientific Management and the Unions, 1900–1932* (Cambridge, Mass.: Harvard University Press, 1955), and Hugh G. J. Aitken, *Taylorism at Watertown Arsenal* (Cambridge, Mass.: Harvard University Press, 1960). Taylor felt that Emerson was hurting his own efforts in management by distorting the system. Taylor to Morris L. Cooke, January 13, 1910, in Taylor Collection.

[41] Oberlin Smith, comment on Frederick W. Taylor, "Shop Management," ASME, *Trans.*, XXIV (1902–3), 1471.

munication with the men: "Each man should be encouraged to discuss any trouble which he may have, either in the works or outside, with those over him. Men would far rather even be blamed by their bosses, especially if the 'tearing out' has a touch of human nature and feeling in it, than to be passed by day after day without a word, and with no more notice than if they were part of the machinery."[42]

Not all were blind to the contradictions of management science. John T. Hawkins observed that following out the logical implications of Taylor's shop methods and system would result, regretfully, in the emasculation of the "real mechanic." Hawkins looked wistfully back to the "old-time conscientious, faithful, intelligent, finger and brain wise mechanic."[43] Few shared Hawkins' perspicuity—most were content to continue giving lip service to the shop culture ideal while adapting and experimenting with various management systems, the only logical result of which was the destruction of that culture.

Rationalization was by its very nature bureaucratic rather than entrepreneurial. James M. Dodge, in a "history" of shop management delivered before the ASME in 1906, stated that the one idea which was characteristic of all such systems was the reduction of the importance of one-man management.[44] H. L. Gantt developed this idea consciously, and paradoxically he was even more interested in the welfare of the individual worker than Taylor. Gantt proposed that manufacturers were afraid of the "system" because it "ultimately makes the proper running of a shop independent of any particular man."[45] Results indicated, Gantt observed, that "SEVERAL MEN WHEN HEARTILY CO-OPERATING, even if of everyday caliber, can accomplish what would be next to impossible for any one man even of exceptional ability."[46] Taylor constantly scored

42 ASME, *Trans.*, XVI (1894–95), 856–903.

43 Comment on F. W. Taylor, "On the Art of Cutting Metals," ASME, *Trans.*, XXVIII (1906), 290–91.

44 "A History of the Introduction of a System of Shop Management," ASME, *Trans.*, XXVII (1905–6), 723. Dodge, a close associate of Taylor's, both business and social, is credited by Charles Piez, *James Mapes Dodge* (n.p., 1916), p. 18, with humanizing Taylor's concepts.

45 Gantt, comment on Taylor, ASME, *Trans.*, XVI (1894–95), 884.

46 Gantt, comment on Taylor, ASME, *Trans.*, XXVIII (1906), 57.

the "unevenness" of management and noted that most shops he had seen were "under-officered." "Functional" management, as Taylor called his system, provided for "each man in the management [to be] confined to the performance of a single leading function." He was also fond of saying that shop bosses were unwilling to take the systematic methods of the office to the shop. The theme of bureaucratization is so strong that it would perhaps be more accurate to say that scientific management was merely one aspect of a broad, general trend toward rationalization and bureaucratization of life in America in the late nineteenth and early twentieth centuries. Taylor and his followers, like other rising technical elites in this period, might have been interested in the heroic exertions of the fabled John Henry, but they would have thrilled at the sight of forty disciplined workmen running their steam riveters at the highest practical speed.[47]

It is not surprising that management engineering became a field and a would-be profession itself. A few leaders like Taylor, Gantt, and Halsey gathered around them bands of disciples and acted within the role of consulting engineers. They did not find an easy road to instant success in this role.[48] Owners and superintendents employed them but interfered with the complete implementation of the systems, resenting, perhaps, the destruction of their roles as unique individuals. Taylor, in desperation, was forced to start his own model company in which he could display his plan in its entirety, unmolested by employers' demands for quick solutions to specific production snags. The heavily entrepreneurial bias of the shop culture elite which controlled the ASME precluded its remaining the major forum for the complete and broad development of management engineering, and such organizations as the Taylor Society were formed to fill this need. Nevertheless, the ASME was not totally unreceptive to the favorable publicity and stature which Taylorism and other scientific management plans were giving the profession of mechanical engineering. Taylor was elected president in

---

47 F. W. Taylor, "Shop Management," ASME, *Trans.*, XXIV (1902–3), 1340–91.

48 L. P. Alford, *Henry Laurence Gantt, Leader in Industry* (New York: American Society of Mechanical Engineers, 1934).

1906, and the new secretary of the ASME, Calvin W. Rice, was known to be active in management engineering circles. By this time (1906–8), however, the ASME had ceased to be the major forum for management questions. The centers of discussion were moving elsewhere, and efficiency engineering became in the progressive era a fetish, often exploited for political purposes.[49]

The business community absorbed and used that part of scientific management in Taylorism which it found acceptable and useful, and the promise of an elite corps of engineers masterminding the systems of industrial production did not find immediate expression. Much of what actually happened in terms of rationalized mass production methods (such as the Ford automobile plants) took place almost entirely outside the confines of "scientific management" or Taylorism. Control of business enterprise would remain in many cases in the hands of uniquely powerful individuals who ran their organizations personally and with little apparent regard for system; in those industries where corporate styles of management were clearly observed, engineers often did not fit into the management roles created by the system.[50]

If scientific management failed to achieve power as a movement, it was perhaps because its leaders were men who held mutually contradictory opinions regarding the relative efficiency of systematic impersonal and unsystematic personal management. Men like Taylor, Towne, and Halsey had grown up through the shop culture system and had become part of its elite. Even as they promulgated systems of management which could have no other effect than to destroy shop culture as they had known it, they gave passionate lip service to the ideal it represented. Perhaps the most curious figure is Taylor himself, giant of the movement, who mercilessly attacked the personal,

---

[49] Morris Cooke advised Taylor that his emphasis on his own complete system, no part of which could be left out, would mean that "the business world will decide against you." Cooke to Taylor, October 7, 1907, in Taylor Collection.

[50] Morris Cooke, again acting in an advisory role to the older man, suggested to Taylor that what was needed were many men trained in the essentials of management techniques to handle the routine work, since "every proposition to be organized is not as complicated as a machine shop." Cooke to Taylor, December 26, 1906, in Taylor Collection.

entrepreneurial, and unsystematic management of the machine shops, who was also one of the most enthusiastic defenders of the shop, with its deep, moving, personal experience of one man in contact with another, boss to workman, as the best education for life a man could get.

As it existed among the mechanical engineers, scientific management may have been seen as a new role for the elite engineers that would replace the shop leadership roles that were being eliminated and absorbed by modern corporate management. It was, like the innovative role in machine and engine design of the earlier period, an opportunity for engineers to shape the methods of production. Shop culture was a source of scientific management figures, and the movement may have been a bridge between roles for some elite engineers. Engineering educators were slow to get behind scientific management, and, when they did in the early twentieth century, it was because it could be connected with the conceptions of social engineering then current among those who were trying to raise the status of engineering in society.

In the two decades from 1840 to 1860, men from the machine-shop culture passed the examinations for and filled positions in the Engineer Corps of the United States Navy. Since they were men of broad experience, gentlemen, and not very numerous, they were allowed to share the privileges of rank with the regular naval officers and were not, before 1860, put under any disadvantage of status or role. It was a respectable and honorable career, and it furnished some of the first mechanical engineers with jobs. The demand for engineers created by the outbreak of the Civil War in 1861, however, altered this informal accommodation and forced both engineer and regular officers to attempt to institutionalize the relationships. It has already been shown that after 1861 large numbers of stationary engineers and other varieties of engine greasers and mechanics had to be commissioned to fill the rapidly increasing need for engineers on the steam vessels of the United States Navy.

By January 4, 1862, William H. Shock, naval engineer, wrote to Benjamin F. Isherwood, Engineer-in-Chief of the Navy, that nearly all the qualified engineers in the Philadelphia area had been signed up and that he suggested looking in Pittsburgh and Wheeling for unemployed marine engineers.[1] John Ericsson, inventor of the *Monitor*, tried throughout the winter of 1863 to find "first-class steam engineers" for the naval positions but found none; he wrote to Isherwood in February that all those he contacted refused to serve the government because of "inadequate pay, compared with the remuneration now offered by companies and private individuals."[2] By February, 1863, standards had been lowered very much and almost anyone could receive a commission as an "acting assistant engineer." The engineer in charge of the *Sangamon*, at war off Newport News, Virginia, complained to Isherwood about his third assistant engineer who had apparently never had "any experience with

---

[1] Entry 970, Vol. 1, RG 19, Bureau of Ships, NA.
[2] February 2, 1863, Entry 970, Vol. 1, RG 19, Bureau of Ships, NA.

Machine shop at the Campbell and Zell Co., Boiler Works,
Baltimore, Md., *ca.* 1900.

any kind of machinery."[3] Such cheapening of a once elite corps led the regular officers to treat all naval engineers as if they resembled the worst, denying them the previously shared privileges of rank. This, as much as the pay, may have led many mechanical engineers to stay out of the service, particularly from 1862 to 1865.

The dozens of letters sent to Isherwood in which the naval engineers complained of slights perpetrated by the sailing officers underscore the contempt which the regular navy men felt for the new breed of engineers. A sample of the engineers' comments includes: "the unpleasant feeling which seems to exist with the sailing officers against us," being assigned sleeping quarters "infested with vermin and . . . extremely warm," and "young officers of the line" took every opportunity to interfere with the engineers' work.[4] Even the regular engineers with prewar service were not immune to such ill treatment, and they too complained to the Chief Engineer of the Navy.

The most common complaint was that the engineers were not permitted the use of the wardroom for mess and sleeping quarters. Typical was the complaint of B. Edward Chassaing, who asked the captain of the ship to which he had been assigned for permission to use a particular room as his quarters. The captain became irate and ordered him out, saying it was to be a chart room. It became, in fact, the captain's bath room. The captain found out that Chassaing had complained to his Engineer Corps superiors and refused liberty to all the engineers for two months in retaliation. Chassaing complained: "[I ask] not only my just rights, but seek to increase the comforts of my brother engineers, by making room in other quarters for them . . . [my room] was unfit for any officer . . . unhealthy . . . devoid of all comfort . . . [and] in one corner of the part of the ship allotted for negroes. . . . It would be impossible for me to

---

3 E. A. C. DuPlaine to Isherwood, February 17, 1863, Entry 970, Vol. 1, RG 19, Bureau of Ships, NA.

4 William H. Fuller, D. M. Grune, E. R. Arnold, and James Butterworth, 3rd Assistant Engineers, U.S.S. *Susquehanna*, to Isherwood, August 1, 1861; Elisha Henderson, U.S.S. *Connecticut*, to Isherwood, February 25, 1862; both in Entry 970, Vol. 1, RG 19, Bureau of Ships, NA. E. J. Brooks, 1st Assistant Engineer, U.S.S. *Richmond*, to Isherwood, February 28, 1865, Entry 970, Vol. 2, RG 19, Bureau of Ships, NA.

relate all the annoyances the engineers as a body and individually on board this ship have met with."[5]

The squabbling became so intense that Secretary of the Navy Gideon Welles decided to clarify the situation by declaring (in an order of September 11, 1863) that "the senior engineer on a vessel had a right to the wardroom, and that the other engineers were to be allowed to mess there for the present by courtesy."[6] This did not end the trouble, however, and the steady stream of complaints kept pouring in, particularly from prewar regular engineers who found that the caliber of the newly appointed assistant engineers was affecting their own status.

The problem, which had not become serious until the elite corps of mechanical engineers was overrun with scores of engine drivers, was that the exact status in terms of rank of the naval engineers was very uncertain. Before the pressures of wartime, the questions of rank had not been crucial to the efficient running of the ship or to survival. But the new monitor-class vessel introduced in 1862–63 had unbelievably little space for crew and officers, making it inevitable that someone had to sleep under the boiler rather than in the cramped wardroom. A high percentage of such complaints came from monitor-class vessels. It was in such vessels also that engineers felt that their special competence was particularly needed—after all, one of their own, John Ericsson, had invented the craft and persuaded the government to use it. A hazy definition of rank between engineer and regular officers under these conditions was a serious matter.

The Navy Department recognized this by the winter of 1863. Under an order of March 13, 1863, relative rank, such as second assistant engineer equals ensign, was spelled out in detail, but engineer officers were still not appointed directly by the President and thus lacked the status of commissioned officers. This replaced the rather unclear law of 1859, which defined rank in terms of "First Asst. Engrs. *next after* lieutenants," and

5 Chassaing to Chief Engineer W. W. W. Wood, August 7, 1861, Entry 970, Vol. 1, RG 19, Bureau of Ships, NA.

6 Copied by F. D. Stedman, 2d Assistant Engineer, U.S.S. *Nipsic*, to Welles, May 15, 1865, Entry 970, Vol. 2, RG 19, Bureau of Ships, NA.

so on. In the period following the end of the Civil War, much adjustment took place, mostly as a result of the shrinking demands for naval vessels and personnel. In 1866 an act of Congress placed the commissions of the corps in presidential hands, but "lack of room on shipboard" was given as a reason for not improving the living and eating accommodations. A year later, in 1867, another act tried to clarify naval engineer status further but resulted in an invidious distinction between line and staff officers, causing some engineers, once professors at the Naval Academy, to be forced to serve under former pupils. This act satisfied few engineers, and a rash of resignations followed. In 1871 a bill defining absolute rank for engineers passed the House of Representatives but was amended in the Senate to a redefinition of "relative" rank, which served until the late nineties when a major reorganization took place.[7] *Army and Navy Journal* noted in 1882 that relative rank was meaningless except as between different services. Why should one class of officers (the engineers) be denied absolute rank since they had "equal professional, educational and social claims."[8] What the engineers were protesting was not the bureaucratic nature of the United States Navy—they were protesting that they as a class of officer did not have the rights or duties to which they ought to be entitled.

The naval engineers, even the high-ranking ones, did not for the most part reject the bureaucratism of the Navy. Reaction to an 1862 bill which would have changed the designation of chief engineer to master engineer illustrates this well. William H. Shock wrote to Benjamin F. Isherwood in early 1862 that the bill would do the infamous disservice of taking "away our Title." Shock went on, "To call any portion of the Chief Eng^{ers} Master Engineers is giving them a title without meaning and offering an indignity they do not deserve."[9] George Sewell,

7 J. W. King to George M. Robeson, December 7, 1869, Entry 963, Vol. 1, RG 19, Letters to Secretary of the Navy, NA; Bennett, *Steam Navy*, pp. 606, 617–18. Admiral David D. Porter wrote in 1866 that the engineers should be made commissioned officers, and that "no class of officers deserves it more." Copy, letter from Porter to William W. W. Wood, June 18, 1866, in Asa Mattice Papers, Cornell University Archives, Ithaca, New York (hereafter cited as Mattice Papers).

8 "Actual Rank for Naval Engineers," *Army and Navy Journal* (hereafter cited as *A. & N. Jour.*), January 7, 1882, p. 501.

9 January 12, 1862, Entry 970, Vol. 1, RG 19, Bureau of Ships, NA.

engineer of the *Susquehanna* off Port Royal, wrote to Isherwood about the same time saying, "be *d– – – – d* to the *ignominious wretch* who put the idea in his [Senator John Sherman, originator of the bill] head, it is evident that some of our many enemies in the line officers have been at work, it is a *dodge* by which they hoped to *jilt* us out of our present *rank and position.*"[10] One engineer actually noted in his comments that the purpose of the bill was to reduce the pay of engineers.[11] The title change was indeed a dodge, but it was the title change, an obvious demotion in status in the naval bureaucracy from their viewpoint, that bothered the top men in the Naval Engineer Corps.

Another provision of the 1862 bill was to create a new grade of fleet engineer, which no engineer opposed, but also to have men for this rank selected by examination rather than by seniority. The Engineer Corps elite almost to a man opposed this move because it would reflect badly upon the chief engineers (the grade below fleet) "who have hitherto been conceded to have passed the final examination requisite to assert a thorough knowledge of their profession," as Theodore Zeller expressed it in a letter to Isherwood.[12] Other comments by top men in the corps reveal that opposition centered around the idea that appointment of a few super-engineers by examination would create dissension in the corps, destroy an elaborate seniority system, and instantly lower the status of all chief engineers. It was as professional and naval bureaucrats that the engineers made these protests. The entrepreneurially inclined would have welcomed a chance to acquire by fair test of ability a slightly higher position than their fellows.[13]

Much of the controversy in the Navy can be attributed to a fight for power and influence, and it lacked philosophical justification. Such were, apparently, the charges of malfeasance in office laid against Benjamin F. Isherwood in 1869. Appointed

10 January 29, 1862, Entry 970, Vol. 1, RG 19, Bureau of Ships, NA.
11 Theodore Zeller to Isherwood, February 5, 1862, Entry 970, Vol. 1, RG 19, Bureau of Ships, NA.
12 February 12, 1862, Entry 970, Vol. 1, RG 19, Bureau of Ships, NA.
13 William W. W. Wood to Isherwood, February 13, 1862; E. D. Robie to Isherwood, February 1, 1862; William H. Shock to Isherwood, February 13, 1862, all in Entry 970, Vol. 1, RG 19, Bureau of Ships, NA.

Chief Engineer in 1861, when the separate Bureau of Steam Engineering was created (following the break up of the old Bureau of Construction, Equipment and Repair), Isherwood had ruled his little corps with an iron hand and had built a close following, largely by virtue of his vigorous defense of the rights and status of individual engineers. His autocratic methods made him enemies also, some of them in engineering itself. In 1863 the Secretary of the Navy appointed a board of nine civilian engineers to consider "certain enquiries bearing on the professional operations of the Chief of the Bureau of Steam Engineering." Some of the charges against Isherwood were on grounds of technical incompetence, some on grounds of actual dishonesty in the form of rebates received personally from firms given naval business by the bureau. According to one board member, J. Vaughn Merrick, a prominent mechanical engineer, Isherwood had the Secretary of the Navy in his pocket, and the board was called to whitewash the affair. The board rebelled at this task, however, and condemned Isherwood's actions; it made extensive recommendations for change in the Bureau of Steam Engineering. The report was almost completely ignored.[14]

Over the next six years, the rumors about Isherwood continued to circulate. The charges of incompetence were compounded by a series of tests which Isherwood conducted at considerable expense on a revolutionary type of vertical-tube boiler. A second civilian board convened in 1869 and recommended Isherwood's dismissal, which was finally secured. The amazing thing is not that he was dismissed, but that it took more than six years of protest for the opinions of the civilian board to be recognized. Isherwood was involved in a personal power struggle in the naval bureaucracy. He had been loyal to those immediately beneath and above him. In spite of internal

14 J. Vaughn Merrick to William Sellers, March 12, 1869, and enclosed memo in an unidentified hand regarding the Board of Inquiry on Isherwood and Statement by Merrick on the same subject, Subject File "EB," Box 133, RG 45, Naval Records, NA. Fortunately, a thorough study of Isherwood's years as Engineer-in-Chief has recently been published. It reveals that Isherwood angered the Philadelphia tool and machinery makers, Sellers, William Bement, and the Merrick interests, by favoring John Roach of New York in the purchase of machine tools. The Philadelphia men, with friends in Congress, called foul when the tools proved to be defective and obsolete. Edward William Sloan, *Benjamin Franklin Isherwood, Naval Engineer: The Years as Engineer in Chief, 1861–1869* (Annapolis, Md.: U.S. Naval Institute, 1965), p. 216.

squabbles, the naval bureaucracy was not about to allow him to be dismissed on the recommendation of an outside civilian board, even though that board had on it the cream of American mechanical engineers. Officially and unofficially, the Navy remained loyal to the man and his memory.[15]

In addition to disputes over the structure of the naval bureaucracy, the Civil War created serious disagreements and conflicts over the introduction of new technology. Alban C. Stimers was a dedicated young engineer in the corps who was determined to remake the American Navy into an ironclad, steam-propelled fighting machine.[16] 1861 found him working on the Stevens Battery, twenty years in the building, a *cause célèbre* of the engineering community. A huge floating ironclad fort covered with revolving gun turrets, the battery was begun in the 1840's, alternately worked on and abandoned by the government, and finally scrapped before completion in the 1870's when it would have been obsolete anyway. In the 1850's and even in the Civil War, American mechanical engineers almost to a man thought that the Stevens Battery would revolutionize coast defenses, and over the years they excoriated the government and the Navy for not pushing for its completion. Stimers was working on the battery as chief engineer of the project in 1861 and was sending glowing reports to Isherwood. He had received the assignment after complaining of his regular superior officer's inclinations to disregard his advice on engineering matters.[17]

When the government began work on the *Monitor*, a new, smaller type of ironclad designed by John Ericsson, Stimers was selected to superintend the building of the vessel. While doing this he wrote with some frequency to Isherwood about problems of the status and rank of naval engineers, expressing the conviction that engineers did not have proper status.[18]

15 Merrick to W. Sellers, March 12, 1869, in NA.; Bennett, *Steam Navy*, p. 609.

16 Stimers was supremely ambitious and an empire builder. Sloan, *Isherwood*, pp. 68–71.

17 Stimers to Isherwood, August 18, December 2, 1861, Entry 970, Vol. 1, RG 19, Bureau of Ships, NA.

18 For example, see Stimers to Isherwood, February 11, 1862, Entry 970, Vol. 1, RG 19, Bureau of Ships, NA.

When he personally directed the running of the radically new vessel to its historic bout with the *Merrimac* off Hampton Roads, Stimers was miffed when the credit all went to the commander of the vessel. Despite the *Monitor's* success and the subsequent action of the government in ordering a fleet of them built in a crash program, many regular naval officers, especially those with emotional commitments to sailing ships, resisted the efficient use and employment of the new craft. Fearing, no doubt, the complete takeover of the Navy by the ironclad steam warship, the running of which was within the competence of naval engineers and not of regular officers, these sailing officers, who represented many of the top wartime command positions, fought the innovation.[19]

An example of this, which led to vast repercussions in the Navy, was the failure of Admiral Samuel Francis Du Pont to use effectively a fleet of monitors in the unsuccessful attack on Charleston in April, 1863. Du Pont was absolutely opposed to the innovation of steam and iron and defended his actions on the basis that the monitor crafts he had were unseaworthy, in need of extensive repairs after an unsuccessful initial assault, and not reliable generally. Stimers, now general inspector of all ironclad vessels, inspected the ships and, finding them in usable condition (with only slight damage), issued a contradictory report. Du Pont, incensed, brought charges against Stimers on grounds of conduct unbecoming an officer. Stimers denied the worst of the charges (based on certain newspaper reports of his statements) before a court of inquiry and was cleared. The case was, however, clearly one of new technology versus old technology, young Turks attacking old guard, innovation versus conservatism. Even though acquitted, Stimers had been brought to trial and forced to deny allegations that he almost certainly did make while irate at Du Pont's actions. The new navy of the engineers would have a hard fight on its hands to win acceptance.[20]

19 The situation was very complex, however; for example, Isherwood was against the monitor craze which developed after the encounter at Hampton Roads. Sloan, *Isherwood*, p. 67.

20 Bennett, *Steam Navy*, p. 403.

In a bureaucracy, the line of command and authority must be clearly delineated. As tensions rose in the Navy over the status of the Engineer Corps and the introduction of ironclad ships of war on a broad scale, naval engineers discovered daily what their relative rank meant or did not mean. Dozens of cases in which the authority of the chief engineer of a vessel over his men and his department was questioned and refuted by the sailing officer-in-charge flooded Isherwood's office in Washington during the Civil War. Typical was the complaint of John Johnson in 1863, an acting second assistant engineer-in-charge on board the U.S.S. *Teazer*. The captain of the vessel took away his firemen and replaced them with boys who were useless. Later the executive officer and the master mate took the rest of his men up on deck to clean "bright work at the guns." When he complained, Johnson was told that these officers were his superiors and he had no right to complain about them.[21] Engineers-in-charge who resisted such tactics more vigorously were court-martialed. Isherwood was apparently able to give little satisfaction in cases like these, involving as they did the basic system of discipline in the Navy. When relief for specific problems did come it came from sympathetic regular officers, like David D. Porter, who ordered in October, 1864, that "for the better preservation of steam boilers in the Squadron, Commanding Officers are required to allow the engineers three hours' time, at least, to raise steam."[22]

Although tensions eased somewhat after the pressure of war was gone, some engineers continued to run into difficulty with the Navy's discipline and command system. The case that attracted the most publicity in the 1890's was that of Assistant Engineer Danforth, who refused the master-at-arms' request for one of his machinist's services while he was conducting an engine test ordered by the chief engineer of the vessel. Danforth further resisted the naval bureaucracy when he refused to report on his actions to the officer of the deck.

---

[21] October 12, 1863, Entry 970, Vol. 1, RG 19, Bureau of Ships, NA.
[22] General Order No. 9, October 17, 1864, Subject File "EV," Box 142, RG 45, Naval Records, NA.

Tried, Danforth was sentenced to one year at half pay and loss of rank. *American Machinist* wrote several editorials about this abusive treatment of an engineer, but the naval bureaucracy nevertheless won the round.[23]

The obvious solution to these difficulties, to combine the talents of line and engineer officers in the same individual, trained at the Naval Academy, was pursued with intermittent vigor after 1861. Steam was made a part of the curriculum that year at the Academy, and in 1863, Engineer-in-Chief Isherwood urged Secretary of the Navy Gideon Welles to push through Congress a bill to establish a special class of cadet engineer to be educated at the Academy.[24] By 1864 Isherwood reported to Secretary Welles that it was essential to upgrade instruction in steam engineering and practical and theoretical mechanics at the Academy.[25] A year later the Department of Steam Engineering was established, and a building for the department was erected in 1866.

Coming as it did, at the end of the war and when the demand for new personnel was waning, neither of these movements achieved much success. The cadet program, a two-year affair, had just two applicants by 1866; the idea was dropped in 1868 but reinstated in 1871 with a large class of sixteen. Made a four-year course in 1874, the cadet engineer program had some success in the seventies, but was finally abolished in 1882. The Steam Department at the Naval Academy had by that time achieved a solid enough footing that regular Annapolis graduates could receive, if they elected to, special training in engineering. Civilian appointments from the engineering schools and from the shops continued to be made. As Annapolis-graduated midshipmen entered the Engineer Corps the scene was set for the logical end to the troubles of the corps. It was, in 1899, simply absorbed into the regular Navy. The transition was made

---

23 "Engineers in the Navy," *Amer. Mach.*, December 1, 1892, p. 8; "The Danforth Case," *Amer. Mach.*, December 8, 1892, p. 8; "Responsibility without Control," *Amer. Mach.*, January 5, 1893, p. 8. See also a letter from Asa Mattice to Gustav Kaemerling, October 18, 1896, in Mattice Papers.

24 Bennett, *Steam Navy*, pp. 654–56.

25 November 26, 1864, Entry 963, Vol. 1, RG 19, Letters to the Secretary, NA.

simpler by a general exodus from the corps over the three decades from 1865 to 1895 of the older group of non-Annapolis engineers.[26]

One cause for this exodus was the decreased chances for promotion due to the ever-shrinking size of the corps itself in the two decades after 1865. From a peak of strength of 2,277 engineers in 1865, the Navy dropped to a low of 173 by 1896. In 1865 a steam frigate of 1,000 horsepower had nine engineers; in 1896 an armored steam cruiser of 17,000 horsepower had only five.[27] This does not necessarily mean that the 1896 cruiser was undermanned. Naval steam engines had reached the height of perfection by the 1890's and probably required far less attention than did the engines of 1864, even though they were far more complex. Such developments did, however, have the effect of reducing the possibilities for careers in this field. An act of 1882 which fixed the total number of assistant engineers at one hundred had the practical effect of stopping promotion altogether in the lower grades.[28] The leading engineers within the corps opposed these changes vigorously, suggesting that efficiency in wartime would be seriously impaired by peacetime cutbacks. In 1884, the Secretary of the Navy's report indicated, to the *American Machinist*'s chagrin, that there were forty assistant engineers who had been in that grade for fifteen to seventeen years without promotion.[29]

It appears that naval engineers were given incentive to leave the corps by the nature of their assignments in the interwar period. Dozens of them were given assignments as professors at engineering schools, which, we have seen, led to mass resignations. Also, "special assignments" were made to machine shops, engine works, and other typical employers of mechanical engi-

[26] Bennett, *Steam Navy*, pp. 655–76; A Cadet on Leave, "Injustice to Cadet Engineers," *A. & N. Jour.*, September 25, 1880, p. 148; Philadelphia *Evening Telegram* of June 18, quoted in *A & N. Jour.*, June 28, 1873, p. 731; a pamphlet distributed among naval engineers quoted in *A & N. Jour.*, January 28, 1872, p. 571. For discrimination against cadet engineers at the Academy and elsewhere, see drafts of letters prepared but not dated or signed, in Mattice Papers.

[27] "Queer Doings in the Navy," *Scientific Machinist*, July 1, 1896, p. 4.

[28] Bennett, *Steam Navy*, p. 751.

[29] "Stagnation of Promotion among Naval Engineers," *Amer. Mach.*, January 12, 1884, p. 3.

neers. Ostensibly, such assignments were for the purpose of supervising tests and engineering work in which the Navy had an interest. Subjectively, it was a simple method of inducing naval engineers to resign to take civilian jobs. Wartime conditions were the only conditions which could possibly create mass demand within the naval bureaucracy for competent and innovative mechanical engineers, and some stimulating work had to be found for them during periods of peace. There was, of course, a hard-core group of naval engineers who believed strongly in the corps and its mission and resisted the trend toward making the corps smaller or abandoning it; but by the 1890's this was a small group indeed. By 1890 the *American Machinist* reported that engineer-officers were resigning at a rate which the editors felt threatened the efficiency of the corps.[30] By 1892, of the 113 living graduates of the Naval Academy engineering course, 38 per cent had resigned to take positions in civil life. If the engineer officers who had promotion open to them were resigning at that rate, one can imagine that the rate for the non-Annapolis engineers was much higher.[31] In a sense naval engineering and the mechanical engineering training at the Naval Academy were performing the kind of function for mechanical engineering that the United States Military Academy at West Point performed for civil engineering.

The engineers who remained with the corps to the end in 1899 developed a cogent argument to support their claims for special status. Some of the more prominent ex-naval engineers echoed the argument. Ira N. Hollis was quoted in the *Army and Navy Journal* in 1897 as saying that "the modern ship is a machine. . . . All the problems on a modern battleship are engineering in their nature, and there is no problem which cannot be solved by the man whose early education has been largely in mechanics and engineering."[32] Another writer, speaking of the line officers, said in 1892, "It is hard for them to see

---

30 "Engineers of the Navy," *Amer. Mach.*, March 13, 1890, p. 8. Mortimer E. Cooley received encouragement to resign his commission to go into teaching. Cooley, *Scientific Blacksmith*, p. 59.

31 "Engineers of the Navy," *Amer. Mach.*, December 22, 1892, pp. 7–8.

32 Hollis's article in *Atlantic Monthly*, quoted in *A. & N. Jour.*, August 28, 1897, p. 962.

that the modern warship is a great fighting machine, and that its success must depend largely upon the engineers . . . who manage the machinery with which she is fitted."[33] *Scientific Machinist* in 1896 reported that the naval regime was tied to a system "in which the machine and the machinist, the engine and the engineer had no place."[34] Technical skill, especially that of the mechanical engineer, was essential to the waging and winning of modern war, or so went the argument.

The line officers put up a last-ditch fight, charging that the engineers were "non-combatants," thus should not enjoy command rank, and that as a class, engineer officers were not fit to be officers. Captain Robley D. Evans put this case forward vigorously in an article in the *North American Review*, widely reprinted in engineering journals for the purpose of criticism. Evans claimed that the engineers wanted to assume titles "which describe another class of officers [and to assume] authority to command and discipline their men," when in fact they did not use the power and authority they already had. Admiral John G. Walker took the position that engineers took no initiative in battle. Lieutenant S. A. Staunton scored engineers for opposing military discipline and trying to bring the methods of industry to ships of war. According to Staunton, they wanted to benefit from greater rank and title without submitting to navy discipline. In short, the engineers were rocking the naval bureaucracy and destroying discipline with their demands for greater status. According to the line, steam had not changed materially the elements necessary for proper command of a ship of war. Discipline and judgment, not technical skill, were the keys to command of the modern Navy, as they had been for the old.[35]

The naval engineers who chose to remain in the corps fought the line in several ways. Through their friends in Congress they

33 *Amer. Mach.*, June 9, 1892, p. 8.

34 "Our Defenses, Our Navy and Naval Men," *Scientific Machinist*, January 1, 1896, p. 4. See also Gustav Kaemerling to Asa Mattice, August 11, 1896, in Mattice Papers.

35 "The Engineer in Naval Warfare," quoting articles in *North American Review* by Robley D. Evans, John G. Walker, and S. A. Staunton, *A. & N. Jour.*, December 5, 1896, p. 236. See also copy of letter from George W. Melville to George Quick, December 10, 1896, Mattice Papers, regarding strategy for combating the image created by the *Review* article.

introduced over the years a number of bills to give them greater status, which were widely supported by engineering journals and by engineering societies of all types, but which did not pass.[36] Another method was to organize a pressure group, the American Society of Naval Engineers, founded in 1889. At their meetings they paid homage to the first-generation giants of the corps like Isherwood and Charles H. Haswell.[37] Although, ostensibly, they formed for social purposes, organizations like the national society and similar local societies resulted in increased pressure for reform of the status of the naval engineer in the 1890's, culminating, to the joy of some and the chagrin of others, in the absorption of the corps in 1899.

Naval engineers, at least when they became ex-naval engineers, played an important role in the American Society of Mechanical Engineers. A few like E. D. Leavitt and George Westinghouse went directly from their positions as engineers in the Civil War Navy to the shop world and made their reputations there. Far more took the route of teaching in the technical schools, especially those who remained in the service or entered after the war. Robert H. Thurston, William F. Durand, Ira N. Hollis, and M. E. Cooley were but a few of the better known. Most of the ex-engineers thus entered the bureaucratic framework of engineering education, even though they had left the Engineer Corps partly out of disgust for naval bureaucracy. They left mainly because there was no opportunity in the naval bureaucracy; they created one of their own in which they could play the major role. It would be fair to say that the naval engineers were a vocal and active group, both those who left the corps to go into education and industry and those who stayed with it to the end, and that they exercised far more power and influence than their mere numbers would suggest.[38]

36 "The Fight of the U.S. Naval Engineer Corps," *Amer. Mach.*, December 19, 1895, p. 1008; *A. & N. Jour.*, March 14, 1896, p. 506.

37 Editorial, *A. & N. Jour.*, January 19, 1889, p. 410; "The Naval Engineers' Dinner," *A. & N. Jour.*, December 24, 1892, p. 297.

38 George W. Melville, "Engineering in the United States Navy, Its Personnel and Material," ASME, *Trans.*, XXI (1899–1900), 139–55. The engineering societies' help was requested in the fight. See undated drafts of an appeal to the AIME, ASCE, and ASME to send representatives to lobby for naval engineering reforms, in the Mattice Papers.

If the naval engineers worked within what was essentially a bureaucratic tradition, then one would expect them to have little respect for the traditions of shop culture, to wish to avoid identification with the mechanic-machinist role, and to stress their role as "Engineers." This was precisely the case. William Shock of the Engineer Corps was instrumental in 1879–80 in securing the abolition of the grade of machinist in the Navy. He asked this on the grounds that the tendency was to employ machinists to do work which should be left to those who were up on the latest advances in the "science of steam engineering." Most machinists were incompetent and untrustworthy, according to the report Shock made to the Secretary of the Navy in 1879. Their ignorance and carelessness had caused need for very costly repairs. Shock suggested and got as a substitute the employment of blacksmiths and finishers, known as "engine room artizans," at a much lower salary than the machinists had been getting.[39] The message is clear. The engineer wanted a clear gap between himself and any class of engine room denizens with whom he might be confused. Secondly, blacksmiths could not be assigned engineers' duties in a pinch, as machinists had been by the commanding officers of the line in the past.

In the sixty years of its existence as a separate entity, the Engineer Corps of the United States Navy led a fitful and uncertain life. Although they came into head-on conflict with naval bureaucracy, the engineers were nevertheless within the same tradition and they certainly carried it with them to the outside world and particularly to engineering education when they left the corps. It was not a case of the engineers opposing bureaucracy per se; it was rather a case of their opposing a bureaucracy unresponsive to their demands for the technical improvement of the Navy, for greater status within the hierarchy for themselves, and for complete recognition as professionals by the line officers.

The nonentrepreneurial experience of many of the profes-

<hr />

[39] Shock to Secretary Thompson, October 16, 1879; Shock to Thompson, November 1, 1880; Shock to William E. Chandler, July 17, 1882; all in Entry 962, Vol. 2, RG 19, Letters to the Secretary, NA.

sional naval mechanical engineers is particularly significant when one considers their influence on engineering education. It is not surprising that the first graduates of the engineering schools were considered deficient in business training and overly scientific in outlook. In the battle between school and shop cultures, naval engineering provided school culture with some of its strongest adherents, who, in turn, helped shape what it became.

American mechanical engineers did not, before 1910, seriously concern themselves with the problem of developing and adopting a formal code of ethics for the profession. The civil engineers, with their high percentage of consultants, had much concern for this problem and so, oddly, did the electrical engineers. A code of ethics was suggested at various times in the technical journals and even in meetings of the ASME, but the subject was never followed up.

American Engineer, always at the fringe of mainline opinion in mechanical engineering, made an appeal in its early editorial period for some formal agreement on professional ethics among engineers. The journal suggested, for example, that engineers should not attack each other in public, more professional courtesy should be shown among engineers, and the public's demand for half-truths and quick solutions to complex problems had to be resisted.[1] *American Machinist*, before 1906, almost never broached the subject, and it was the real spokesman for predominant opinion in the field of mechanical engineering. Within the ASME, two notable attempts were made to advance the idea. Oberlin Smith suggested the need for a formal code in his 1890 presidential address, and William F. Durfee asked in 1886 for a code which would define the duty of a young engineer to his chief, relation and duties of one engineer to another, duty which an engineer owes to his employer, and the duty which the profession owes to society at large. Smith, as we have seen, advocated a much more high-and-mighty role for the engineer than most ASME members would respond to, publically or otherwise, and Durfee was actually using the idea of a code of ethics to control the young college graduate, not the untrained shop engineer. Durfee went on to say that young engineering school graduates tended to undermine their chiefs, were wont to spell assistant with a big *A*

---

1 "Professional Ethics among Engineers," *Amer. Engr.* (Chicago), I (August, 1880), 119.

Tool room of the National Cash Register Corp., 1904 (Library of
Congress photo).

and chief with a small c. Clearly, there was no widely supported call for a code of ethics that would protect the engineer or the public.[2]

Doctors, lawyers, and a few other professionals did by this time have rather well-defined codes of ethics, even though these were sometimes on a state rather than a national level. Sometimes these codes were formal like the physicians' Hippocratic oath, and sometimes they were informal and unwritten, as frequently the ethics of the bar were. It is a fact, however, that professionals of this type are in greater need, for their own protection and the public's, of such formal codes since they are constantly in situations in which their honor and professional reputation could be compromised. Clearly, some sort of guidelines were necessary, such as the privileged character of secrets confided to a doctor or clergyman, for example. The client-practitioner relationship was a meaningful concept in the traditional professions of medicine, law, and the ministry, but could have little meaning in an occupation like mechanical engineering, where perhaps 3 per cent were independent consultants with this kind of relationship before them.

In response to a few demands for a code of ethics, engineers developed the idea that since the actions of the engineer were checked at every point by the immutable laws of God and nature, there was no possibility for undetected malfeasance. This concept was developed in conjunction with the idea that the traditional professions, despite their fancy codes of ethics, had far greater opportunity to cheat and defraud the individual client and the public. Professor J. B. Johnson expressed this clearly:

Men in other professions may blunder or play false with more or less impunity. Thus the lawyer may advocate a bad cause without losing caste; a physician may blunder at will, but his mistakes are soon buried out of sight; a minister may advocate what he no longer believes himself, and feel that the cause justifies his course; but the mistakes of the engineer are quick to find him out and to proclaim his incompetence. He is the one professional man who is obliged to be right.[3]

2 Discussion of Woodward, "Training of a Dynamic Engineer," ASME, *Trans.*, VII (1885–86), 769–71, 776–80.

3 "Two Kinds of Education for Engineers," *Addresses*, ed. Waddell and Harrington, p. 30.

265

Victor C. Alderson, of Armour Institute of Technology, shared similar views. To Alderson, the engineer was bound by his scientific data, and "departure from these data means failure." Other professional men were subject to mere human laws and notions of right and wrong, and few citizens or clients had the technical competence to judge them anyway. Alderson concluded, however, that "a mere tyro can recognize a poor roadbed, defective machinery, or a dangerous bridge."[4] This last comment, thrown in as a minor point by Alderson, is a key to the understanding of the nature of this ethical noncode.

It was true not only that God or nature passed judgment on the engineer's work, but also that the average layman could tell at once whether the engineering work was sound. If there was one thing the average American considered himself, it was a mechanic, and as such he was qualified to judge engineering design and correctness. Thus, not only did the mechanical engineer deal with men who knew what they wanted and might well be engineers themselves, but he faced firm notions of what was desired on the part of ordinary businessmen and manufacturers with no engineering training. Hedged by God and nature on one side and on the other by knowledgeable employers, the engineer was given little room to commit breaches of "ethics."

By way of contrast to that of the mechanical engineers, the civil engineers' organization and profession at large after about 1893 developed some concern for a code of ethics. Even though the movement to create a code was defeated within the organization, there was a large segment who were in favor of it.[5] Likewise, the electrical engineers became concerned about a code of ethics early in the twentieth century. In 1906 Schuyler Skaats Wheeler of the AIEE proposed an extensive code, which elicited a great deal of favorable discussion.[6] Some of the provisions were unrealistic, as the ever-practical *American Machinist* was quick to point out. Referring to a provision that

---

[4] "Ethics of the Engineering Profession," *Railroad Digest*, reprinted in *Amer. Mach.*, May 9, 1901, pp. 521–23.

[5] J. A. L. Waddell, "Address to the Graduating Class of the School of Engineering at the University of Kansas," *Addresses*, ed. Waddell and Harrington, p. 369; "The Regulation of Engineering Practice by a Code of Ethics," pp. 45–62.

[6] "Engineering Ethics," *Amer. Mach.*, July 5, 1906, pp. 8–11.

no electrical engineer should take a position vacated by another unless he was satisfied that the latter engineer left the position voluntarily and for proper reasons, the mechanical engineering journal suggested that "no code of ethics can be enforced which conflicts with the will of the employer." The journal went on: "It is not to be believed for a moment that any electrical manufacturing company will ever find the slightest difficulty in filling a vacant position because of this provision." Noting a clause in the proposed electrical engineers' code which demanded that the engineer quit when he was overruled by superiors in strictly engineering matters, *American Machinist* declared: "The employer is the final judge of the manner in which his money shall be spent. He is usually more or less of an engineer himself, and frequently a very competent one." Commenting generally on the proposed code, the journal summarized: "They [the electrical engineers] have forgotten that while lawyers work for clients, engineers work for employers. . . . The mechanical engineer and electrical engineer are . . . essentially employees, and we believe that any attempt to lay down a comprehensive code of ethics to apply to those professions must be abortive."[7]

After 1906, some interest in the idea of a code for mechanical engineers appeared, but its supporters were few and nothing was done about it. Members of the business profession and the eminently practical profession realized that the role of the mechanical engineer precluded an enforceable and workable code of ethics based on the client-practitioner relationship peculiar to other professions. This meant that mechanical engineers did not develop (as professionals) as much concern for larger public questions as did their sister professions of civil and electrical engineering.

Mechanical engineers as a class tended to agree that the average politician at the national, state, and local level was not only corrupt, but incompetent to deal with the problems of running a modern technological society. *American Machinist* categorically declared in 1895 that "in this most mechanical age mechanical fitness and ability is more scarce than it should

7 "Engineering and a Code of Ethics," *Amer. Mach.*, July 25, 1907, p. 141.

be among men of wealth and power who have control of all great affairs."[8] Oberlin Smith was convinced that the only way to reform the patent office was to "kill all the present Congressmen and put members of the engineering societies in their places." Another engineer told the ASME that the average congressman was influenced by the motive of pleasing the lower class, not inventors and engineers, and he recommended civil service reform for the patent office troubles. Commenting on the same question, Professor Thomas Egleston felt it would be sufficient to "increase the average intelligence of the ordinary Congressman."[9] *American Machinist* noted in 1890 that it was perfectly feasible from an engineering standpoint to bridge the Hudson River, but that the project was held up in Albany by political squabbling.[10] A logical question: What could the engineer do to improve the conditions under which public issues were settled?

There is a style of address found most commonly at graduations, Fourth of July celebrations, inaugurations, dedications, which is not the exclusive property of mechanical engineers or any other single group. It stresses the mission of the group being spoken to, which is to make a better world for them and their children to live in. Mechanical engineers were fond of making this sort of address, describing engineering science itself as the religion of salvation from the woes the lawyers, financiers, politicians, had laid upon it over the centuries. The engineer, man of the twentieth century, would strike away divisions between men by destroying the major obstacle to human brotherhood—the unequal distribution of wealth and creature comforts. They would do this not by socialistic methods, but by creating more wealth to go around through engineering science. This was a convenient kind of social responsibility since it meant that the engineer need only practice his profession diligently and the millennium would straightaway come about.

Robert Thurston's inaugural address as first president of the ASME is full of this kind of large social role for the mechanical

8 "Is Ignorance Ever Innocence?" *Amer. Mach.*, October 31, 1895, p. 868.
9 "Proceedings of the New York Meeting," ASME, *Trans.*, VI (1884–85), 19, 23–24.
10 "Engineering Progress and the Politicians," *Amer. Mach.*, May 8, 1890, p. 10.

engineer.[11] *American Engineer*, in its editorial period in the early 1880's, spilled a lot of ink over such exhortations. The journal felt that the engineer must concern himself with "the social, the religious, the living problem of the present day and of America . . . relating to the just reduction and distribution of wealth." Though afraid that the engineer might cut a sorry figure as a political economist, *American Engineer* could point to Herbert Spencer, the English sociologist and economist who had indeed been an engineer before he took to social philosophy.[12] As examples of the engineer's failure to have influence on legislation and government the Chicago journal noted the failure to reinstate government testing programs. Scoring the engineer's failure to use the power he had in the community, the editors declared: "It is directly within the meaning and the power,—we think, within the function and the duty—of the government to appropriate funds for the solution of such important problems, which effect the whole community and the lives of the people."[13]

By the turn of the century, educators and publicists were suggesting that the engineer was especially fit to deal with the problems of society. As H. G. Prout, editor of the *Railroad Gazette*, said to the graduating class of Stevens in 1899: "For some generations society has agreed to leave to the engineer the care of friction and gravity, but natural depravity has been left to the ministers, lawyers, editors, teachers, the mothers of families, to any one, in fact, but the engineer, and this is where society makes a mistake. The best corrector of human depravity is the engineer, because he must deal with conditions, while others may deal with symptoms."[14] This, William H. Bryan told the Washington University Association in 1903, was the age of the mechanical engineer. He must be "a good citizen, taking part in the civic movements of his day, and throwing whatever light he can upon municipal, state, and national engineering problems. He must render this service at a personal

---

11 "Our Progress in Mechanical Engineering," ASME, *Trans.*, II (1881), 451; see also Thurston, Letter to the Editor, *Amer. Engr.* (Chicago), January 19, 1883, p. 36.

12 "Engineers and Political Economy," *Amer. Engr.* (Chicago), December 22, 1882, pp. 283–84.

13 Editorial, *Amer. Engr.* (Chicago), September 17, 1885, p. 111.

14 "The Engineer and His Country," *Wisc. Engr.*, III (January, 1899), 77–78.

sacrifice, if need be. In such questions as water-supply, rapid transit, street improvement, lighting and drainage, smoke abatement, the engineer as a citizen, whether in or out of office, has a public duty."[15]

When, from 1906 until the early 1920's, the idea of public responsibility as the duty of the engineer was developed and flourished,[16] it was not the mechanical engineers who led the movement, or at least not the old shop-culture elite. The real impetus came particularly from the civil engineers, whose work experience led them into contact with public problems and government, and from electrical engineers, who were anxious to achieve greater professional status and saw public service as a means to that end. Within the framework of the progressive era and its ideology, the movement became confused with scientific management and conservation ideas, finally taking a form that was far outside the boundaries set by the ASME at least.

The ideology of the business profession prevented any more than a passing flirtation with the social responsibility movement that took shape in engineering circles after 1908. Even such innovations as better ventilation and lighting in the shops had to be presented not primarily as humanitarian measures but in terms of the higher productivity and efficiency that would result. The ASME had its collection of papers on the "large role" for engineers just as did the other societies. When H. F. J. Porter in 1905 called for the realization "of Ideals in Industrial Engineering," and in essence suggested a massive effort by engineers in concert with other civic leaders to raise every workman to the level of a useful citizen, W. S. Rogers confessed to "an emphatic coolness of interest respecting the so-called ideals in industrial engineering. They make the most valuable sort of

15 "The Mechanical Engineer, His Duties, Responsibilities, and Opportunities," *Engineer* (N.Y.), April 15, 1903, p. 301.

16 Layton, "American Engineering Profession," pp. 80ff., develops this theme extensively, indicating an acceptance of the idea by the ASME with which I cannot agree. His recent articles emphasize more the idea that certain figures outside engineering, like Thorstein Veblen, thought there was such a mass movement among engineers. See "Veblen and the Engineers," *American Quarterly*, XIV (Spring, 1962), 64–72, and "Frederick Haynes Newell and the Revolt of the Engineers," *Midcontinent American Studies Journal*, IV (Fall, 1962), 17–26.

reading advertisements ever given to a magazine artist, but I have little use for the whole affair." Rogers felt that the movement aimed at telling the worker how to live, and he would have none of it. John T. Hawkins expressed the opinion that Porter's position was simply unrealistic and did not directly approach the practical problem of lessening the power of unions. In fact, very little interest was shown in Porter's proposals, except for the two critical remarks noted.[17]

The need for at least some lip service to the responsibility issue was made apparent at the dedication of the union Engineering Societies Building in 1907. Arthur T. Hadley, president of Yale University, told the assembled engineers: "If the engineer and the lawyer accept positions as servants, simply putting their technical knowledge at the disposal of merchant, journalist or politician who will pay the highest price for it, it is not merely a confession of inferiority—it is a dereliction of public duty."[18] President Theodore Roosevelt's 1908 White House Conference on Conservation led to joint meetings in New York of delegations from the various engineering societies to discuss conservation problems. The idea was picked up by the scientific management clique within the ASME, and papers on the "large role" for engineers began to receive a more favorable hearing in the ASME. For one thing, the new secretary, Calvin W. Rice, was in favor of the idea and he had much control over the selection of papers.[19]

The landmark paper was given by Morris L. Cooke before the ASME in 1908, entitled "The Engineer and the People."[20] Overnight, it seemed, the idea had become respectable. Frederick W. Taylor immediately registered his approval of Cooke's talk and noted Calvin Rice's recent part in arranging the White House conference. Yale's Arthur T. Hadley suggested that mechanical engineering would not reach a high level of professionalism unless Cooke's ideas were appreciated. Oberlin Smith,

17 "The Realization of Ideals in Industrial Engineering," ASME, *Trans.*, XXVII (1905–6), 343–72.

18 "Account of the Dedication of the Engineering Societies Building," ASME, *Trans.*, XXIX (1907), 58.

19 "Regular Monthly Meetings, April 14, Conservation of Our Natural Resources," ASME, *Trans.*, XXX (1908), 11–25.

20 ASME, *Trans.*, XXX (1908), 619–37.

rewarded after twenty years of campaigning, applauded Cooke's expression of what he had been talking about all along.[21] The idea did not long remain bounded by separate engineering disciplines. It was almost by its very nature a pan-engineering movement, which may never have reached the rank and file of society membership. It was promoted by men, like Morris L. Cooke, who later went into government service and were not seeking entrepreneurial careers, who were engineers with a capital "E."[22]

In effect, Cooke and others were arguing for a truly professional role for the mechanical engineer. Cooke suggested that as a technician the engineer was now recognized as a professional, whereas he was not recognized as a public figure. He cited the activities of the American Medical Association, which was conducting campaigns for health education of the public. When the ASME was too slow in following his suggestions for professional development, Cooke lost interest in the organization and turned his energies to more profitable use elsewhere.[23]

Local and regional engineering societies, containing as they did representatives from all branches of engineering, and having a community tie in common, were much more active in supporting the idea of public responsibility, particularly before 1908 when the idea caught on with the national societies. The Chicago-based Western Society of Engineers began serious discussion of the question of public responsibility and duty when in 1891 L. E. Cooley spoke on "The Modern Spirit of the Engineering Profession." The doctor and lawyer, according to Cooley, advocated measures for the public good which fell

21 Cooke had sent advance copies of his paper around to prominent ASME members for comment. He wrote to F. W. Taylor that he feared many men might be frightened away from the idea unless the "right kind of people" endorsed it in open meeting. Cooke to Taylor, October 24, 1908, in Taylor Collection. See also Taylor to Cooke, October 15, 1908, ibid., in which Taylor anticipated the question in most engineers' minds, "Just what is to be done." Taylor's advocacy in the meeting helped to allay Cooke's fears that his idea would be sidetracked by "older and more conservative members." Cooke to Taylor, November 3, 1908, ibid.

22 For Cooke's later activities see Kenneth E. Trombley, The Life and Times of a Happy Liberal: A Biography of Morris Llewellyn Cooke (New York: Harper and Brothers, 1954).

23 Ibid., pp. 16, 44–70. Cooke was enamored of the close relationship which the faculty of the University of Wisconsin had with political reform in that state and hoped to see the engineer develop such a role nationally. Cooke to Taylor, February 23, 1909, in Taylor Collection.

within their competence and so should the engineer. He likened society to a great organic machine and proposed that "sooner or later the engineer will stand at the throttle of this machine and the world will want to know what manner of man he is."[24] Along these lines, Robert Gillham of the Engineers Club of Kansas City outlined in 1893 a broad role for his organization in improving city sanitation, refuse removal, lighting, and paving.[25] The suggestions made to local societies would have shocked the staid ASME membership. In 1895, for example, the retiring president of the Technical Society of the Pacific Coast told his members: "But if money be taken by equitable taxation from those who have by accident or natural qualifications been placed in a position to accumulate beyond their immediate personal needs and desires, then the more that be taken and put into circulation through such channels as the construction of public works, the greater must be the general prosperity of the nation."[26] A member of the Engineers Club of Cincinnati proposed in 1901 that the engineers should organize mutual home-building efforts composed of landowners, material men, merchants, and unemployed laborers and mechanics. He hoped the membership would not think him utopian.[27] Some struggled to keep the movement within the acceptable limits of the free enterprise tradition, such as president W. A. Layman of the Engineer's Club of St. Louis in 1906. Although it was "essentially the business of the engineer to serve the public constructively," Layman insisted that "I do not invite a departure from a policy of conservatism; I rather invite a campaign of creative *work*."[28] One thing stands out—differences between types of engineers were erased by advocacy of public service and responsibility for the engineer.

It is not entirely clear just how the national engineering societies so suddenly plunged into concern for and action in favor

---

24 AES, *Jour.*, X (1891), 63–68.

25 "Work for Our Engineers' Club," AES, *Jour.*, XII (1893), 305–13.

26 C. E. Grunsky, "The Industrial Problem of the Pacific Coast," AES, *Jour.*, XIV (1895), 381.

27 James A. Stewart, "A Plan to Utilize Unemployed Labor," AES, *Jour.*, XXVII (1901), 81–84.

28 "The Engineer's Club in Its Relation to the Future St. Louis," AES, *Jour.*, XXXIX (1907), 64–66.

of a public duty for the engineer. Almost certainly they jumped on board an unstoppable bandwagon which combined the elements of efficiency engineering, the popularization of conservation ideas, so-called progressive political changes at the national, state, and local levels, and a general call for "uplift." Nothing else could explain the sudden and complete about-face on the issue of the relatively conservative ASME in 1908. It was at least partly an indigenous growth with the engineering societies west of Ohio, which challenged the national societies for power and influence. It is not surprising that Minard L. Holman, a St. Louis municipal engineer, was elected president of the ASME in 1908. Not only was he the first president from the West (excluding Ohio), but he was active in the area of public service and conservation. His presidential address was devoted to "The Conservation Idea as Applied to the American Society of Mechanical Engineers." The engineer, Holman told the ASME, must be a social innovator, but within limits. His methods must be consistent with the present level of profits and the profit system, and the engineer must certainly not get very far ahead of the times.[29]

Nevertheless, further explanation of the sudden upsurge of interest in both ethics and a sense of public responsibility about 1906 to 1908 is needed. Actually, the suddenness is more apparent than real; the technical journals, always quick to exploit a circulation-building crusade, made it appear that there was vast interest in the topic while in fact such interest did not transform the ASME, at least not overnight. The fact that in this period engineers were asked to participate actively in conferences on such public issues as conservation and national efficiency must have helped to give the impression that the entire profession was awakening to a sense of public responsibility. A national efficiency mania was getting under way in these years, to culminate in public interest in the railroad hearings of 1910. Thus, conservation and efficiency engineering, and through them the idea of a larger public role for engineers, became in the years from 1906 to 1908 highly newsworthy, generating interest as well as reflecting it through magazines

<hr>

[29] ASME, *Trans.*, XXX (1908), 577–617.

and newspapers. The mechanical engineering press was definitely one of the imitators, not an innovator, in developing the concept of a larger role for the engineer.[30]

In practice the ASME continued its old policy of nonaffiliation and noninvolvement. In 1908 the council declined an invitation to join the Conservation League of America and in 1910 decided not to associate with either the League of Good Roads or the Good Roads Association. These organizations were actually concerned with subject areas related to engineering, as was the subject of smoke abatement, which suffered a similar neglect within the ASME. Morris Cooke enlisted the aid of 150 ASME members in petitioning the meetings committee to organize a conference on the abatement of smoke in cities and industrial areas. The conference was to include talks by engineers and by physicists, chemists, architects, physicians, and others outside engineering and cooperation with the American Civic Federation, and clearly reflected the large-role concept. The council, speaking for the meetings committee, turned down the request without comment.[31]

Nothing better illustrates the conservative, shop-culture conditioned approach to these new ideas than what happened to Frederick W. Taylor in 1910. In 1906 Taylor was president of the ASME and at the height of his power within the organization. By late 1910, railroad men among the ASME membership were aghast at the testimony of Harrington Emerson in the Eastern Rate case, suggesting that the railroads could save one million dollars per day by using scientific management procedures. Taylor had perfected his own theories on management and prepared a paper on them which he hoped to present before the ASME. He got a very calculated cold shoulder from the

[30] The idea of radicalism and insurgency appealed to Cooke along with the notion of being a progressive, all terms in current vogue in the political spectrum. See letter from Cooke to F. W. Taylor, March 12, 1909. The influence of outside figures is seen in Cooke's description of an interview with Ray Stannard Baker, which he reported to Taylor: "Mr. Baker is tremendously interested in 'Insurgency' wherever he finds it, and he says that he looks upon you and your group as the insurgents among engineers." Cooke to Taylor, December 10, 1910, in Taylor Collection.

[31] October 13, 1908, October 11, 1910, and November 12, 1910, ASME Council Minutes, note these rejections. On smoke abatement see entry for June 2, 1909, ASME Council Minutes. That Cooke was something of a promoter, trying to push the ASME, is indicated by his comment to Taylor, "I don't know anything about 'smoke' myself so I am looking to you and others to keep me out of trouble." Cooke to Taylor, December 6, 1908, in Taylor Collection.

meetings committee; when he raised a ruckus by writing to members personally about this, the secretary wisely shifted the burden of decision to the membership via a circular letter concerning preferences for papers. Taylor was told "the membership does not want papers of this kind." He fumed but eventually published his "Principles of Scientific Management" elsewhere. The ASME leadership clearly did not like the direction in which concern for management, efficiency, and conservation was taking them and the society.[32]

When certain engineers after World War I tried to take the engineer ahead of his time, it was not done within the framework of the basically conservative ASME. Management engineering was a distinct contribution which mechanical engineering made to American society in the period from 1880 to 1910. In the case of public responsibility, the notion was not a contribution by the engineers; it was instead an idea absorbed from outside. The training and background of some engineers made it possible for them to accept a large role and even to carry this role to its logical end of actual public service. The ASME, however, was unfit by its tradition and heritage for leadership in such a movement. Because the organization failed to take a leading part in this movement, engineers who wanted action turned to the formation of new organizations, which cut across engineering lines and which promised action on social and professional issues.

The idea that mechanical engineers developed a broad sense of public responsibility and took an active part in the organization and administration of society is not a correct one. It is an impression gained from ideological evidence produced by a few publicists and activists in the profession. School culture found the idea useful in the elevation of the status of the college-trained engineer, and the educators quickly incorporated it into their ideology of professionalism. Shop culture often demurred, finding the concept of the engineer as technocrat incompatible with the image of the engineer-entrepreneur. Whatever the ideology expressed by both groups, neither had implemented it in practice by 1910.

[32] See correspondence Taylor to M. L. Cooke, September 26, October 6, October 17, December 2 and 10, 1910, in Taylor Collection.

The professionalization of the American mechanical engineer was complicated by the conflict of two distinct cultures for control of the process of selection, training, and socialization of the young mechanical engineer. These two cultures, shop and school, differed in both aim and method. Most of the internal differences over the question of professional behavior can be related to this conflict.

Shop culture was a personal affair. Interpersonal relations in a small, experimental machine or engine shop in the period from 1820 to 1890 were close and intimate. Although large numbers of those who ultimately did become mechanical engineers had social and economic status which aided them, nevertheless, apprenticeship was a meaningful experience. Mechanical engineers did "take to" certain apprentices in the shop and try to educate them to be mechanical engineers. That most of the apprentices they picked were men of their own class does not suggest collusion as much as it suggests that early mechanical engineers (who constituted a self-recognized elite) found it more pleasant and practical to work closely with those they could trust, converse with intelligently, and—not to be discounted—see socially. Nonetheless, a strong myth of upward mobility became attached to the shop culture idea, a myth which was perpetuated by the elite of mechanical engineering.

It was possible, so read the myth, for one to begin as a poor but honest and hardworking mechanic in the shop, thence to raise oneself to the level of the mechanical engineer and even beyond, to proprietor of one's own shop. The ASME, founded in 1880, made so much of the few men who had risen from blacksmith or mechanic to world-famous engineers (such as John Fritz and John Edson Sweet) that one is tempted to conclude that such a rise was not at all typical. Most elite mechanical engineers (those in the ASME or otherwise recognized by the profession) came from middle- or upper-class backgrounds. Of the very top engineers, nearly all seem to have come from upper-class backgrounds and frequently were related to each other by blood or marriage. These individuals defined what a mechanical engineer was (for the United States) and thus set the definition to include themselves but to exclude others informally. By the social and economic status they brought to the profession of mechanical engineering, they made it a gentlemen's calling. Corollary to this was a conception of themselves as an elite and a belief that mechanical engineering should be kept an elite occupation. This did not conflict with the mobility myth because only those would rise who were natural aristocrats.

By contrast, school culture was impersonal and stressed the importance of such external factors as examination scores in judging a man's worth. Educators were interested in raising large numbers of average boys to be average engineers, not in the perpetuation of a self-recognized elite. School culture believed in the high school, not in the shop, as the source of future engineers and discounted the value and importance of shop culture as an educating and socializing force. Formal education was seen by the educators as the coming instrument of social and economic mobility. By the 1920's, candidates for degrees in mechanical engineering were coming mainly from the lower middle class and were receiving unexpected gains in status from their studies. The educators were trying to create large numbers of engineers to fill the demand created by changing industry, and they were trying to broaden the field of engineering.

Shop culture was centered around the small machine shop, frequently a partnership, whose success required the entrepreneurial spirit. School culture, on the other hand, oriented its training programs toward the large bureaucratic corporation, which was indeed one of the principal markets for the educator's product after 1890. Here in microcosm can be seen the conflict of the old and new forms of business and industrial organization. Shop culture itself was modified as the shop changed to larger corporate manufacturing enterprises. We can thus say that the engineering educators recognized the specialization of the decision-making function and that the shop culture elite did not. Scientific management, as the mechanical engineers reacted to it and created it, illustrates this difference. Educators recognized, more than did shop men, that decision-making was increasingly being separated from engineering work in industrial processes. Those who made decisions in large corporations were men who left engineering proper and became administrators. Hence the bureaucratic orientation of their approach to training the engineer. Shop culture advocates reacted to the changes differently: one of their responses was to create scientific management, which was a way for the engineer *as engineer* to retain decision-making powers and to cope with the largeness, impersonality, and corporateness which was coming to characterize the old shop world they knew. Scientific management provided a role for the elite engineer; significantly, it did not take over American industry (as a movement) as the engineers planned. Nevertheless, it was an innovative function which the shop culture elite (who constituted most of the leadership in scientific management) employed to retain the foothold in industry which they had so long held. Innovation was not the function of the school, which had to keep potential demand for their product in mind at all times. Thus engineering educators were slow to get behind scientific management and did not truly embrace it until the early twentieth century when the entire society did so.

In spite of paying lip service to eclectic, intuitive, "practical" methods, the shop elite were among the first to use precise experimentation and science in the shop. Often one finds that the most vigorous defenders of shop culture, the intuitive element, and "finger knowledge" were simultaneously the innovators in introducing science to the shop. The ultimate reference point, however, is that the key to shop culture was *applied* science. Engineering educators put emphasis on pure science and on calculus and higher mathematics as universal tools which trained the mind to do any task. Intuition was linked by them with superstition and the Dark Ages. Consistency was valued over pragmatic eclecticism.

The meaning of these differences over the use of science is clear when one surveys the reaction of the two groups to rationalization in all forms and the way in which the educators were able to characterize shop culture as an institution opposed to science and thus to progress. Shop culture opposed rationalization only when it was to be imposed by a centrally established authority with arbitrary power, or when such rationalization offered change without significant gain in efficiency and production. Partly this reaction was entrepreneurial and partly it was related to their conception of engineering as a science of getting the most for the dollar. School culture used the element of rationalization and standardization as a wedge to get its graduates into good jobs. It succeeded in the newer industries in which large numbers of scientifically competent (not necessarily brilliant) engineers were needed. The educators favored centrally established standards because they felt they would be thus uniformly and bureaucratically administered and probably because they felt that they would have a hand in determining those standards. Scientifically determined and imposed standards would in their view give order to the irrational market place. Shop culture was ideologically committed to the market place, refused to take action on standards, and opposed all centrally imposed standards, even though it privately worked to create them by business pressure.

Feeling themselves already members of an elite profession, assured of status by birth, entrepreneurial activity, or engineering skill, the shop elite was not interested in the substantive questions of professionalization. They were even opposed to some of the elements of professionalism since they tended to downgrade entrepreneurial status and to separate it from engineering status. Any arbitrary qualifications for admission to the mechanical engineering fraternity were also opposed by the shop men because of the difficult-to-define quality that held them together in professional association—class and pre-existing status—and because formal qualifications might exclude some of the most eminent among them who had not bothered in many cases to get a formal education. Professionalism might mean that exit from mechanical engineering would have to be controlled, again something that might tend to exclude businessmen who had once been engineering innovators and who had much weight in the profession. The ASME, tightly controlled by the shop elite, performed quite adequate social and intellectual functions and was not asked to indulge in a scramble for greater status for the engineer.

Naturally the engineering educators wanted to change the rules of the game. Their students, unless they already had the proper qualifications, did not go into the best eastern shops and did not receive the deference which the shop elite engineers did. By strongly pushing the issue of professionalization and making the requirements for engineering practice formal, the college graduate had an immediate advantage over the self-trained shop engineer. The college-trained were put at a disadvantage in doing this in that the professional association was controlled by the antiprofessional forces. Gradually, however, aided by some compromises with the shop elite on curriculum and by new, expanding markets for their graduates, the educators acquired a voice in the councils of the ASME and began to achieve a few of their goals. The conflict between shop and school began in the late 1860's when the first engineering schools for mechanical engineers were set up, intensified from 1880 to 1890—the formative period of the ASME—was followed by a period of self-examination and compromise from 1890 to 1905, and ended with the ascendancy of the school forces after 1905. By this last date the engineering schools were in more or less full control of the training process of mechanical engineers.

It is ironic that the groups which opposed professionalism among mechanical engineers were the same individuals who had first set the seal of profession on their work. They did oppose it in the latter nineteenth century because it seemed symptomatic of changes that were destroying the small, personal industrial community centered around the machine shop, which they all knew and loved. Because they controlled the professional organization and wielded power in engineering circles, they were able to resist effectively for some time the claims of the educational complex to exclusive domination over the training and professionalization of the American mechanical engineer.

Manuscript Collections

AMERICAN PHILOSOPHICAL SOCIETY, PHILADELPHIA, PENNSYLVANIA
    PEALE-SELLERS PAPERS
AMERICAN SOCIETY OF MECHANICAL ENGINEERS, NEW YORK, NEW YORK
    COUNCIL OF THE ASME, MINUTES
CORNELL UNIVERSITY ARCHIVES, ITHACA, NEW YORK
    BOARD OF TRUSTEES, MINUTES
    BOARD OF TRUSTEES, PAPERS
    EXECUTIVE COMMITTEE OF THE BOARD OF TRUSTEES, MINUTES
    ASA MATTICE PAPERS
    SIBLEY COLLEGE PAPERS
    ROBERT H. THURSTON PAPERS
    ANDREW DICKSON WHITE PAPERS
NATIONAL ARCHIVES, WASHINGTON, D.C.
    BUREAU OF SHIPS, RECORD GROUP 19
    LETTERS TO SECRETARY OF THE NAVY, RECORD GROUP 19
    NAVAL RECORDS, RECORD GROUP 45
    RECORDS AND MINUTES OF EXAMINING BOARDS, RECORD GROUP 24
SMITHSONIAN INSTITUTION, WASHINGTON, D.C.
    ALEXANDER LYMAN HOLLEY LETTERBOOKS
    WILLIAM RICH HUTTON COLLECTION
STEVENS INSTITUTE OF TECHNOLOGY, LIBRARY, HOBOKEN, NEW JERSEY
    FREDERICK WINSLOW TAYLOR COLLECTION

Books

AITKEN, HUGH G. J. *Taylorism at Watertown Arsenal.* CAMBRIDGE, MASS.: HARVARD UNIVERSITY PRESS, 1960.
ALFORD, L. P. *Henry Laurence Gantt, Leader in Industry.* NEW YORK: AMERICAN SOCIETY OF MECHANICAL ENGINEERS, 1934.
AMERICAN SOCIETY OF MECHANICAL ENGINEERS. *Diamond Jubilee Book of Facts.* NEW YORK: AMERICAN SOCIETY OF MECHANICAL ENGINEERS, 1955.
———. *Fifty Years Progress in Management, 1910–1960.* NEW YORK: AMERICAN SOCIETY OF MECHANICAL ENGINEERS, 1960.
———. *First Catalogue of the American Society of Mechanical Engineers.* NEW YORK: AMERICAN SOCIETY OF MECHANICAL ENGINEERS, 1880.
———. *Fred J. Miller.* NEW YORK: AMERICAN SOCIETY OF MECHANICAL ENGINEERS, 1941.
BALTZELL, E. DIGBY. *Philadelphia Gentlemen.* GLENCOE, ILL.: FREE PRESS, 1958.

BENDIX, REINHARD. *Work and Authority in Industry.* NEW YORK: JOHN
WILEY & Co., 1956.

BENNETT, FRANK M. *The Steam Navy of the United States.* PITTSBURGH:
WARREN & Co., 1896.

BISHOP, J. LEANDER. *A History of American Manufactures.* 2 VOLS.
PHILADELPHIA: E. YOUNG AND Co., 1861.

BISHOP, MORRIS. *A History of Cornell.* ITHACA, N.Y.: CORNELL UNIVERSITY
PRESS, 1962.

BROEHL, WAYNE G., JR. *Precision Valley.* ENGLEWOOD CLIFFS, N.J.:
PRENTICE-HALL, INC., 1959.

BURR, HENRY L. *Education in the Early Navy.* PHILADELPHIA: No PUB-
LISHER, 1939.

BURSTALL, AUBREY F. *A History of Mechanical Engineering.* LONDON:
FABER AND FABER, 1963.

CALHOUN, DANIEL H. *The American Civil Engineer.* CAMBRIDGE, MASS.:
M.I.T. PRESS, 1960.

————. *Professional Lives in America.* CAMBRIDGE, MASS.: HARVARD UNI-
VERSITY PRESS, 1965.

CAPLOW, THEODORE. *The Sociology of Work.* MINNEAPOLIS: UNIVERSITY OF
MINNESOTA PRESS, 1954.

CHANDLER, ALFRED D., JR. *The Railroads—The Nation's First Big Business.*
NEW YORK: HARCOURT, BRACE, AND WORLD, 1965.

COLVIN, FRED H. *60 Years with Men and Machines: An Autobiography.*
NEW YORK: McGRAW-HILL, 1947.

COOLEY, MORTIMER E. *Scientific Blacksmith: The Autobiography of Mortimer E. Cooley.* NEW YORK: AMERICAN SOCIETY OF MECHANICAL
ENGINEERS, 1947.

COPLEY, FRANK BARKLEY. *Frederick W. Taylor, Father of Scientific Management.* 2 VOLS. NEW YORK: HARPER AND BROTHERS, 1923.

*Description of the Collections of Scientific Appliances Instituted for the
Study of Mechanical Art in the Workshops of the Imperial Technical
School of Moscow.* MOSCOW: W. GAUTIER, 1876.

DRURY, HORACE BOOKWALTER. *Scientific Management.* 3RD ED. NEW YORK:
COLUMBIA UNIVERSITY PRESS, 1922.

DUPREE, A. HUNTER. *Science in the Federal Government.* CAMBRIDGE, MASS.:
HARVARD UNIVERSITY PRESS, 1957.

DURAND, WILLIAM F. *Adventures in the Navy, in Education, Science, and
in War.* NEW YORK: AMERICAN SOCIETY OF MECHANICAL ENGINEERS,
1953.

————. *Robert Henry Thurston.* NEW YORK: AMERICAN SOCIETY OF ME-
CHANICAL ENGINEERS, 1929.

ENGINEERING ASSOCIATION OF THE SOUTH. *Selections from Papers Presented
during the Fiscal Year Nov. 21, 1889, to Nov. 13, 1890.* NASHVILLE,
TENN.: UNIVERSITY PRESS, 1891.

ENGINEERS' SOCIETY OF WESTERN PENNSYLVANIA. *Pittsburgh: Commemorat-*

ing the *Fiftieth Anniversary of the Engineers' Society of Western Pennsylvania*. PITTSBURGH: THE SOCIETY, 1930.

FERGUSON, EUGENE S. (ED.). *Early Engineering Reminiscences (1815–1840) of George Escol Sellers*. WASHINGTON, D.C.: SMITHSONIAN INSTITUTION, 1965.

FRITZ, JOHN. *The Autobiography of John Fritz*. NEW YORK: JOHN WILEY & SONS, 1912.

GIBB, GEORGE SWEET. *The Saco-Lowell Shops*. CAMBRIDGE, MASS.: HARVARD UNIVERSITY PRESS, 1950.

GODDARD, DWIGHT. *Eminent Engineers*. NEW YORK: THE DERRY-COLLARD Co., 1906.

HABER, SAMUEL. *Efficiency and Uplift: Scientific Management in the Progressive Era, 1890–1920*. CHICAGO: UNIVERSITY OF CHICAGO PRESS, 1964.

HALSEY, FREDERICK A. *The Metric Fallacy* AND SAMUEL S. DALE. *The Metric Failure in the Textile Industry*. TWO BOOKS PUBLISHED AS ONE. NEW YORK: VAN NOSTRAND, 1904.

HITCHCOCK, EMBURY A. *My 50 Years in Engineering*. CALDWELL, IDAHO: CAXTON PRINTERS, LTD., 1939.

HUTTON, FREDERICK REMSEN. *A History of the American Society of Mechanical Engineers from 1880 to 1915*. NEW YORK: AMERICAN SOCIETY OF MECHANICAL ENGINEERS, 1915.

JOHNSON, WALTER R. *Address Introductory to a Course of Lectures on Mechanics and Natural Philosophy, Delivered before the Franklin Institute, Philadelphia, November 19, 1828*. BOSTON: HIRAM TUPPER, 1829.

KIRBY, RICHARD SHELTON (ED.). *Inventors and Engineers of Old New Haven*. NEW HAVEN, CONN.: NEW HAVEN COLONY HISTORICAL SOCIETY, 1939.

KNOLL, H. B. *The Story of Purdue Engineering*. WEST LAFAYETTE, IND.: PURDUE UNIVERSITY PRESS, 1963.

KORNHAUSER, WILLIAM. *Scientists in Industry*. BERKELEY: UNIVERSITY OF CALIFORNIA PRESS, 1962.

LINCOLN, SAMUEL B. *Lockwood Greene: The History of an Engineering Business, 1832–1958*. BRATTLEBORO, VT.: STEPHEN GREENE PRESS, 1960.

LIPSET, SEYMOUR M., AND BENDIX, REINHARD. *Social Mobility in Industrial Society*. BERKELEY: UNIVERSITY OF CALIFORNIA PRESS, 1959.

LOHR, LENOX R. (ED.). *Centennial of Engineering*. CHICAGO: CENTENNIAL OF ENGINEERING, 1953.

MANN, CHARLES R. *A Study of Engineering Education*. NEW YORK: CARNEGIE FOUNDATION FOR THE ADVANCEMENT OF TEACHING (BULLETIN No. 11), 1918.

MARX, LEO. *The Machine in the Garden*. NEW YORK: OXFORD UNIVERSITY PRESS, 1964.

MERTON, ROBERT K. *Social Theory and Social Structure.* GLENCOE, ILL.: FREE PRESS, 1949.

MUMFORD, LEWIS. *Technics and Civilization.* NEW YORK: HARCOURT, BRACE, & CO., 1934.

NADWORNY, MILTON J. *Scientific Management and the Unions, 1900–1932.* CAMBRIDGE, MASS.: HARVARD UNIVERSITY PRESS, 1955.

NAVIN, THOMAS R. *The Whitin Machine Works since 1831.* CAMBRIDGE, MASS.: HARVARD UNIVERSITY PRESS, 1950.

NOSOW, SIGMUND, AND FORM, WILLIAM H. (EDS.). *Man, Work and Society.* NEW YORK: BASIC BOOKS, 1962.

PASSER, HAROLD C. *The Electrical Manufacturers, 1875–1900.* CAMBRIDGE, MASS.: HARVARD UNIVERSITY PRESS, 1953.

PIEZ, CHARLES. *Personal Reminiscenses of James Mapes Dodge.* NO PLACE: NO PUBLISHER, 1916.

PORTER, CHARLES T. *Engineering Reminiscences.* NEW YORK: JOHN WILEY & SONS, 1908.

REDLICH, FRITZ. *History of American Business Leaders.* 2 VOLS. ANN ARBOR, MICH.: EDWARDS BROS., 1940.

RICKETTS, PALMER C. *History of Rensselaer Polytechnic Institute, 1824–1914.* NEW YORK: JOHN WILEY & SONS, 1914.

ROE, JOSEPH W. *English and American Tool Builders.* NEW HAVEN, CONN.: YALE UNIVERSITY PRESS, 1916.

———. *James Hartness.* NEW YORK: AMERICAN SOCIETY OF MECHANICAL ENGINEERS, 1937.

ROGERS, CHARLES BARTON. *Objects and Plan of an Institute of Technology.* 2ND ED. BOSTON: JOHN WILSON AND SON, 1861.

ROLT, L. T. C. *A Short History of Machine Tools.* CAMBRIDGE, MASS.: M.I.T. PRESS, 1965.

ROSS, EARLE D. *Democracy's Colleges.* AMES, IOWA: IOWA STATE COLLEGE PRESS, 1942.

SCHARF, J. THOMAS, AND WESTCOTT, THOMPSON. *History of Philadelphia (1609–1884).* 3 VOLS. PHILADELPHIA: L. H. EVERTS & CO., 1884.

SHAW, RALPH ROBERT. *Engineering Books Available in America Prior to 1830.* NEW YORK: NEW YORK PUBLIC LIBRARY, 1933.

SLOAN, EDWARD WILLIAM. *Benjamin Franklin Isherwood, Naval Engineer: The Years as Engineer in Chief, 1861–1869.* ANNAPOLIS, MD.: U.S. NAVAL INSTITUTE, 1965.

SMITH, ALBERT W. *A Biography of Walter Craig Kerr.* NEW YORK: AMERICAN SOCIETY OF MECHANICAL ENGINEERS, 1927.

———. *John Edson Sweet.* NEW YORK: AMERICAN SOCIETY OF MECHANICAL ENGINEERS, 1925.

STRASSMAN, W. PAUL. *Risk and Technological Innovation.* ITHACA, N.Y.: CORNELL UNIVERSITY PRESS, 1959.

TROMBLEY, KENNETH E. *The Life and Times of a Happy Liberal: A Biog-*

raphy of *Morris Llewellyn Cooke*. NEW YORK: HARPER AND BROTH-
ERS, 1954.

WADDELL, JOHN A. L., AND HARRINGTON, JOHN L. (EDS.). *Addresses to
Engineering Students*. KANSAS CITY: WADDELL AND HARRINGTON,
1911.

WESTINGHOUSE ELECTRIC CORPORATION. *The Engineering Graduate*. PITTS-
BURGH: THE WESTINGHOUSE CORPORATION, 1925.

WICKENDEN, WILLIAM E. *A Comparative Study of Engineering Education
in the United States and in Europe*. LANCASTER, PA.: SOCIETY FOR
THE PROMOTION OF ENGINEERING EDUCATION, 1929.

WIESSENBORN, G. *American Engineering, Illustrated By Large and Detailed
Engravings Embracing Various Branches of Mechanical Art, Sta-
tionary, Marine, River Boat, Screw Propeller, Locomotive, Pumping
and Steam Fire Engines, Rolling and Sugar Mills, Tools, and Iron
Bridges, of the Newest and Most Approved Construction*. NEW
YORK: G. WIESSENBORN, 1861.

WILSON, LOGAN. *The Academic Man*. NEW YORK: OXFORD UNIVERSITY
PRESS, 1942.

WOODBURY, ROBERT S. *History of the Grinding Machine*. CAMBRIDGE,
MASS.: TECHNOLOGY PRESS, 1959.

Unpublished Materials

CALVERT, MONTE A. "AMERICAN TECHNOLOGY AT WORLD FAIRS, 1851–1876."
UNPUBLISHED MASTER'S THESIS, UNIVERSITY OF DELAWARE, 1962.

LAYTON, EDWIN T. "THE AMERICAN ENGINEERING PROFESSION AND THE IDEA
OF SOCIAL RESPONSIBILITY." UNPUBLISHED PH.D. DISSERTATION,
UNIVERSITY OF CALIFORNIA AT LOS ANGELES, 1956.

SINCLAIR, BRUCE. "DELAWARE INDUSTRIES: A SURVEY, 1820–1860." UN-
PUBLISHED RESEARCH REPORT, ELEUTHERIAN MILLS-HAGLEY FOUNDA-
TION, DECEMBER, 1958.

287

INDEX

Alden, George I., 78
Alderson, Victor C.: on code of ethics, 266
Allen, Leicester, 209
Allis-Chalmers Company: training program, 75
American Civic Federation, 275
American Engineer: on engineering education, 54, 80; described, 107, 134; on the founding of the ASME, 111; on engineering as a profession, 156; on status of engineers, 158, 161; on titles, 163–65; on engineering unity, 219–20; on engineers as businessmen, 233–34; appeals for code of ethics, 263; on civic duty of engineers, 269
American Engineers' Association: activities, 107–8
American Institute of Electrical Engineers: founding of, 213–15; membership described, 216–18; concern for code of ethics, 218–19; on engineering unity, 219
American Institute of Mining Engineers: mechanical engineers as members of, 109; founding of, 199, 210–12; lacks interest in professionalism, 213; mentioned, 77, 170
American Machinist: on technical education, 55, 79; on demand for engineers, 59, 60; in defense of shop training, 68; survey on apprenticeship, 72–73; on college admission standards, 76–77; on Sibley College, 89, 95; founding and purposes of, 110; on founding of ASME, 112; on geographical distribution of ASME membership, 119; on change in ASME dues, 122; on quality of papers at ASME meetings, 124–25; on license laws for engineers, 127; on local engineering societies, 134; described, 136–38; on titles, 140, 162–66; on engineering positions, 154; on status of engineers, 157; on use of resources, 169; on role of machinists, 191–92; on social status of mechanics, 194–95; on practice of electrical engineering, 217; on Engineering Societies Building, 222; opposition to civil engineers, 223–24; on scientific management, 235–36, 238; on efficiency of the Engineer Corps of the Navy, 256–57; on code of ethics, 263, 266, 267, 268
American Medical Association, 272
American Railroad Journal: on railway engineering, 14; mechanical engineering department of, 16; on need for steam engineers, 20; on mechanical occupa-

tions, 29–30; on architects, 208–9; mentioned, 135
American Railway Review, 135
American Railway Times, 135
American Society of Civil Engineers: mechanical engineers as members of, 109; and engineering inclusiveness, 200–1; attitude toward ASME, 203; on role and status of civil engineers, 204–7; concern for code of ethics, 207–8; professional orientation of, 208–9; opposes Engineering Societies Building, 222–24; mentioned, 77, 170, 220. See also Civil engineering
American Society of Mechanical Engineers: founding of, 109–11; as an elite organization, 112, 126, 154; qualifications for membership in, 112–14; membership composition, 114–20; government of, 121–25; professional activity of, 125–26, 130–31; on public issues, 126–27; opposes license laws for engineers, 127–28; organizational changes after 1906, 128–30; attitude toward local societies, 133, 134*n*, and government testing commissions, 170, 172; and standardization of engineering terms, 172–73; on standards of machine design, 176–78; controversy over metric system, 179–86; attitude toward stationary engineers, 189–91; and ASCE, 199; assisted by ASCE, 203; relation to AIME, 210; relation to AIEE, 215; and engineering unity, 221; leaders as businessmen, 229; relation to scientific management, 236–43; role of naval engineers in, 259; attitude toward code of ethics, 263–65; attitude toward civic role for engineers, 270–76. See also *American Machinist*
American Society of Naval Engineers: aims of, 259
Apprenticeship: in conflict over engineering training, 72–76; debated by journals, 194; mentioned, 277
Architects: professional position discussed, 157, 208–9
Army and Navy Journal: on relative rank for naval engineers, 249
Ashworth, Daniel: opposed to engineering unity, 221
Association of Engineering Societies: purpose of, 133; on engineering unity, 219–20
Atkinson, Edward, 99

Babcock, George, 119–20
Babcock, John, 45

Polytechnic Institute, 44–45; influenced by Europe, 47, 53–54; Robert H. Thurston on, 47, 104; influence of Morrill Land-Grant Colleges Act on, 47–49; at University of Wisconsin, 48; at Stevens Institute of Technology, 49; at Purdue University, 49–50; and stress on mathematics, 54–55; role of technical schools in, 55; and engineering research, 55–56; as a profession, 57–58; *American Engineer* as a voice for, 137–38; admission standards at schools for, 76–77; degrees, 165–66; business techniques in, 232–34

Educators of mechanical engineers: and criticisms of shop training, 63–65, 77–80; concern for status, 71–72, 158–59; on titles, 165–66; on the metric system, 182–86; on business training, 232

Egleston, Thomas: on government testing commissions, 171–72; defines *engineering*, 226; mentioned, 268

Electrical engineers: professional role of, 213, 215, 217; and code of ethics, 218–19; and engineering unity, 219. *See also* American Institute of Electrical Engineering

Electrical World: on engineering, 215

Elite: concept of, in mechanical engineering, 8; common characteristics of, 11–12; as defenders of shop culture, 70; on training programs for technical graduates, 75; ASME as organization of, 112; views of, on shop structure, 191

Emerson, Harrington: on scientific management, 239; mentioned, 275

Emery, Charles E.: in defense of shop training, 68

Engine-driver occupation: and license laws, 27; relation to mechanical engineering, 27. *See also* Stationary engineering

Engineer: on railway engineering, 15; on mechanical engineering writers, 53; on technical graduates, 61; on engineering education, 80; on engineering periodicals, 135; on the role of the mechanical engineer, 142; on machine design, 173–74; on electrical engineering, 215; opposed to scientific management, 238

Engineer Corps of the United States Navy: as source of mechanical engineers, 19–23; character of, before 1860, 22, 245; officers of, detailed to engineering colleges, 50–51; standards of admission to, 245–46; status of, in Navy, 248–50, 254–55,257–59; elite of corps opposed to rank change in, 250; and training at U.S. Naval Academy, 255–56; efficiency of, 256–57; officers of, criticized, 258

Engineering Association of the South, 68, 134

Engineering News: on engineering inclusiveness, 199; opposed to Engineering Societies Building, 223

Engineering Societies Building, 128, 221–24

Engineer's Club of New York, 221

Engineer's Club of St. Louis: membership composition, 134; and engineering unity, 220; mentioned, 131

Engineers' Society of Western Pennsylvania: membership composition, 134

Ericsson, John: correspondence with Robert H. Thurston, 99; mentioned, 101, 203, 245, 247, 252

Ethics, code of, for engineers: ASCE concern for, 207–8; ASME attitude toward, 263–65; mentioned, 218–19, 263–68

Evans, Oliver, 25–26, 203

Evans, Robley D.: and criticism of naval engineers, 258

Fairbanks and Bancroft Company, 9, 10, 11

Falkenau, Arthur: and Sweet-Morris controversy at Cornell, 90–93, 95; in favor of the metric system, 183

Forney, Matthias N.: and change in ASME dues, 122

Franklin Institute: and technical education, 43; and study of licensing of engineers, 127

Franklin Institute Journal: discussed by *Engineer*, 135

Freeman, John R.: on engineering education, 81

Fritz, John: relation to Robert H. Thurston, 101; mentioned, 277

Fuller, T. H.: on need for engineers, 59

Gantt, Henry L.: on role of mechanical engineers, 141; and scientific management, 237, 240–41

Gardiner, Edward B.: on engineering education at Cornell University, 91–92

General Electric Company: training program, 74–75, 144; as employer of electrical engineers, 217

Gillham, Robert: on civic role of engineers, 273

Globe Nail Company, 59

Good Roads Association, 275

Goss, William F. M.: and engineering education at Purdue University, 50

Hadley, Arthur T.: on civic role of engineers, 271

Halsey, Frederick A.: in defense of shop training, 69–70; and Sweet-Morris controversy at Cornell, 90–93, 95; on factions within the ASME, 125; and organizational changes in ASME, 129; opposed to the metric system, 183; and scientific management, 237–38, 241–42; mentioned, 70, 79

McClellan, William: on status of engineers, 159
McCormick, Cyrus: correspondence with Robert H. Thurston, 99
Machine design: by William Sellers, 173; *Engineer* on, 173–74; *Technologist* on, 174; discussion of, in ASME, 174
Machine shops: described, 5–8; as source of mechanical engineers, 12–13
Machine tools: defined, 4
McNair, F. W.: on engineering education, 55

Magill, Edward H.: as teacher of Robert H. Thurston, 45; mentioned, 96
Magruder, William T.: on machine design, 175
Massachusetts Institute of Technology, 44
Mattice, Asa M., 183
Maxim, Hiram: correspondence with Robert H. Thurston, 99
Mechanic: on technical graduates, 61; mentioned, 33
Mechanical Engineer: on role of mechanics, 191; mentioned, 136
Mechanical Engineering, 137
Mechanical Engineering Teachers Association, 64
Mechanical News: on titles, 165
Mechanics: organizations, 1820–50, 29–32; need for education, 33–37; status of, 37–39, 194–95; lack of professionalism among, 40; position in industrial society, 191–92; as businessmen, 232–34
Mechanics: on government of ASME, 121
Mechanics' Mirror: on status of mechanics, 30–31; on scientific education, 36
Meier, E. D.: on scientific management, 239
Melville, George: correspondence with Robert H. Thurston, 99; on role of the engineer, 143; on status of engineers, 158
Merriam, John C.: as founder of *American Engineer,* 107–8
Merrick, J. Vaughn: opposition to Benjamin F. Isherwood, 251
Metcalf, William: eulogizes Alexander Lyman Holley, 226
Metric system: controversy in ASME, 179–86
Michigan Engineering Society: membership composition, 134
Miller, Fred J.: and change in ASME dues, 122; in favor of the metric system, 183; mentioned, 137
Millwright occupation: as source of mechanical engineers, 25
Mining engineers: and types of employment, 211–12; and correspondence schools, 212; and mining conditions, 212–13; lack of professionalism among, 213
Moore, Lycurgus B.: on early mechanical

organizations, 108; on geographical distribution of ASME membership, 118
Morison, George S.: on mechanical engineering 203; mentioned, 199
Morrill, Justin, 47
Morrill Land-Grant Colleges Act, 47–48
Morris, John L.: and conflict over management of Sibley College, 89–96; Walter C. Kerr's opinion of, 96; mentioned, 87
Mumford, Lewis, 25

National Association of Stationary Engineers, 189–90
Naval engineers: as source of mechanical engineers, 19–23; as college teachers, 50–51; concern for status, 158–59, 248–50, 254–55, 257–59; role in ASME, 259. See also Engineer Corps of the U.S. Navy
Navy Department of the United States: on relative rank for naval officers, 248–49
New York State Mechanic: on education of mechanics, 35; on status of mechanics, 38–39
Nicodemus, William J. L., 48
Northwestern Mechanic: on mechanics' ambitions, 192; on role of mechanics, 193
Norton, Charles, 152

Ohio Mechanics Institute, 44
Ohio State University: and engineering research, 56

Poole, J. Morton, 8–9
Porter, Charles T.: opposed to the metric system, 181; mentioned, 92, 99, 120, 175, 182
Porter, David D.: sympathetic to naval engineers, 254
Porter, H. F. J.: on status of engineers, 157; on civic role of engineers, 270
Powel, Samuel G., 94
Practical Mechanic: on role of mechanics, 193
Pratt and Whitney Company, 12, 176, 177
Pritchett, Henry S., 183
Professionalism: and mechanics' organizations 1820–50, 29, 40; and ASME, 125–26, 130–31; and stationary engineers, 189–90; discussed, 205; of mechanical engineers, 271
Professionalization: defined, xv–xvii; of mechanical engineers, 280–81
Protection of Geneva Mechanics, 39
Prout, H. G.: on importance of engineers, 160; on civic role for engineers, 269
Providence Steam Engine Company, 45
Pupin, Michael, 216
Purdue, John, 49
Purdue University: and engineering education, 49–50; and engineering research, 56

293

THE MECHANICAL ENGINEER IN AMERICA, 1830–1910:
Professional Cultures in Conflict

by Monte A. Calvert

designer: Gerard Valerio
typesetter: Monotype Composition Company, Inc.
typefaces: Weiss Roman
printer: Universal Lithographers, Inc.
paper: 55 lb. Old Forge Offset
binder: Moore & Co., Inc.
cover material: Interlaken AVO #390 and AVI #850